C. Collewet
F. Chaumette
J. Salvi

Visual Servoing by Means of Structured Light

Jordi Pages
C. Collewet
F. Chaumette
J. Salvi

Visual Servoing by Means of Structured Light

Visual-based Robot Task Optimisation from Controlled Light Pattern Projection

VDM Verlag Dr. Müller

Impressum/Imprint (nur für Deutschland/ only for Germany)

Bibliografische Information der Deutschen Nationalbibliothek: Die Deutsche Nationalbibliothek verzeichnet diese Publikation in der Deutschen Nationalbibliografie; detaillierte bibliografische Daten sind im Internet über http://dnb.d-nb.de abrufbar.

Alle in diesem Buch genannten Marken und Produktnamen unterliegen warenzeichen-, marken- oder patentrechtlichem Schutz bzw. sind Warenzeichen oder eingetragene Warenzeichen der jeweiligen Inhaber. Die Wiedergabe von Marken, Produktnamen, Gebrauchsnamen, Handelsnamen, Warenbezeichnungen u.s.w. in diesem Werk berechtigt auch ohne besondere Kennzeichnung nicht zu der Annahme, dass solche Namen im Sinne der Warenzeichen- und Markenschutzgesetzgebung als frei zu betrachten wären und daher von jedermann benutzt werden dürften.

Coverbild: www.purestockx.com

Verlag: VDM Verlag Dr. Müller Aktiengesellschaft & Co. KG
Dudweiler Landstr. 99, 66123 Saarbrücken, Deutschland
Telefon +49 681 9100-698, Telefax +49 681 9100-988, Email: info@vdm-verlag.de
Zugl.: Girona, University of Girona, Diss., 2005

Herstellung in Deutschland:
Schaltungsdienst Lange o.H.G., Berlin
Books on Demand GmbH, Norderstedt
Reha GmbH, Saarbrücken
Amazon Distribution GmbH, Leipzig
ISBN: 978-3-8364-9019-1

Imprint (only for USA, GB)

Bibliographic information published by the Deutsche Nationalbibliothek: The Deutsche Nationalbibliothek lists this publication in the Deutsche Nationalbibliografie; detailed bibliographic data are available in the Internet at http://dnb.d-nb.de.

Any brand names and product names mentioned in this book are subject to trademark, brand or patent protection and are trademarks or registered trademarks of their respective holders. The use of brand names, product names, common names, trade names, product descriptions etc. even without a particular marking in this works is in no way to be construed to mean that such names may be regarded as unrestricted in respect of trademark and brand protection legislation and could thus be used by anyone.

Cover image: www.purestockx.com

Publisher:
VDM Verlag Dr. Müller Aktiengesellschaft & Co. KG
Dudweiler Landstr. 99, 66123 Saarbrücken, Germany
Phone +49 681 9100-698, Fax +49 681 9100-988, Email: info@vdm-publishing.com

Printed in the U.S.A.
Printed in the U.K. by (see last page)
ISBN: 978-3-8364-9019-1

Contents

Chapter 1

Introduction

The confluence of robotics and artificial vision constitutes the framework of this work. This chapter presents a general overview of the topics involved in the work as well as its motivation and objectives. In addition, the outline of the work is presented at the end of the chapter.

1.1 Robots and visual perception

Nowadays two main concepts linked to the term "robot" prevail. One of them, which can be considered as the most extended among people, conceives a robot as a human-like machine which is able to think, sense, act and even feel as humans do. Therefore, a robot is seen as an entity in which both the human anatomy an intelligence are modelled. This definition can be extrapolated to other life forms. Such a romantic concept has been strongly influenced for the vast literary work and filmography produced over the last century. An alternative interpretation of the term robot, and the most realistic one, arises from the scientific community. In this case a robot refers to any machine which is able to fulfill a physical task autonomously or supervised by an operator [Schilling, 1990]. This concept, which probably deceives most science-fiction readers, better fits our current limitations when trying to model and imitate human or animal gaits and, predominantly, when modelling intelligent behaviours.

Surprisingly, the term "robot" was not initially adopted by a scientist but by the Czech writer Karel Capek who first used this word in 1920 in his play "R.U.R." (Rossum's Universal Robots) [Capek, 1970]. In his native language, "robota" refers to a tedious labour. This can be considered the main joint between the literary and the scientific concepts: a robot is intended to perform hazardous, dangerous, heavy or repetitive tasks. Indeed, it has been the main goal of the robots built since the second half of the twentieth century: industrial robots used in manufacturing lines, automotive industry, mine finding, land/underwater/space exploration, manipulation of radioactive materials, etc. In industrial environments, the most typical configuration consists of articulated robots which attempt to mimic the human arm. Such structure provides an efficient solution for applications like assembling, loading, welding or painting. Nevertheless, other type of robot taxonomies are widely used as wheeled and walking robots, which are more suitable for certain applications like navigation, cleaning, transport or even entertainment.

Some authors consider the programmable automates the first generation of robots. However, while programmable automates are only able to know their internal state, robots have the possibility to interact and change its surrounding world by perceiving it and adapting it to its changes. That is why different types of sensors are being increasingly used in robotics such as: tactile sensors, force sensors, proximity sensors, ultrasonic sensors and optical sensors. All these sensors allows a robot to obtain data describing the state of the world analogously to the senses for humans.

Vision is the physical sense consisting of the ability to detect light and interpreting the images formed in our brain in order to *perceive* and understand the environment. Vision is the most important and developed human sense, which allows us to perceive colour, texture, shape and depth. Analogously, computer vision is a subfield of artificial intelligence that investigates how to make computer algorithms which are able to perceive and understand the world through images. Roughly, computer vision tries to emulate the visual perception of the human being from the first stage of light detection till the complex task of understanding what is being perceived. In this case, light is detected by cameras, while the high level tasks of image understanding are processed by using computer algorithms. The goal of achieving a comprehensive perception of the world by computer vision is still far from being attained. The main reason is that the way how the human brain functions is still pretty unknown. Hence, the computer vision community is still handling with low-level problems related to vision like colour and texture perception and intermediate-level problems like motion detection, depth and shape acquisition and object recognition. Another reason which makes hard to emulate the human perception is that we, humans, take profit of our active perception capabilities in order to optimise perceptive tasks. For example, we can converge or diverge the eyes, move the head, change our point of view, etc. The use of these capabilities in order to optimise the perception tasks is also an important research field in computer vision known as *active vision* [Aloimonos *et al.*, 1987].

Among the perception tasks, depth perception is a very important ability for moving in and interacting with the three-dimensional world where we live. It is well known that the human ability of perceiving depth is based on the binocular stereopsis formed by the eyes. The slightly different position of the eyes on the head provokes that an object appears in different horizontal positions in each image provided by each eye. This difference on relative positions, known as disparity, gives a cue about the object's depth. Computer vision tries to copy human stereopsis by using two cameras as if they were two eyes in what is called *passive stereovision* [Hartley and Zisserman, 2000]. An alternative consists in using a single camera and moving it to different known positions for perceiving the scene from multiple points of view. This approach is known as *structure from motion* [Huang and Netravali, 1994]. Furthermore, disparity variations on a sequence of stereo images can be used for rigid motion estimation [Armangué *et al.*, 2003; Perez de la Blanca *et al.*, 2003].

Stereovision is one of the most important topics in computer vision since it allows the three dimensional position of an object point to be obtained from its projective points in the image planes [Faugeras, 1993]. The most difficult problem in stereovision is the determination of homologous points in two images, i.e. determining which pair of projective points represent the same three dimensional point. This problem is known as the *correspondence problem*, which is the main limitation of stereovision since once it is solved all has been already formalised [Faugeras, 1993]. Even if a set of geometrical constraints, known as the *epipolar geometry* [Zhang, 1998], is able to simplify the correspondence problem, it is not a definitive solution. For example, the correspondence problem cannot be solved when observing non-textured objects, when points only appear in one of the images due to a surface occlusion, when points are multiply matched between the images or under adverse lighting conditions.

A solution to the correspondence problem is provided by *active stereovision*. This technique is called active not because it uses active vision but because it is based on changing the lighting of the scene by projecting structured light on it. The use of structured light solves the limitation of passive stereovision when observing non-textured objects. The former structured light techniques were based on projecting simple primitives like a single dot or a single line of light, usually provided by lasers. The advantage of projecting such structured light primitives is that the correspondence problem of the illuminated points in the images is directly solved. Nevertheless, the number of correspondences per image is very small. In order to increase the number of correspondences, structured light patterns were introduced like arrays of dots, stripes, grids or concentric circles. However, with this solution the identification of different pattern regions in the images becomes ambiguous so that the correspondence problem is not directly solved. This fact provoked the emergence of *coded structured light* [Batlle *et al.*, 1998]. In this case, the projected patterns are coded so that each element of the pattern can be unambiguously identified in the images. Therefore, the aim of coded structured light is to robustly obtain a large set of correspondences per image independently of the appearance of the object being illuminated and the ambient lighting conditions.

1.2   From sensing to acting: an approach based on visual servoing and structured light

Humans interact with the environment thanks to a global interpretation of the world. A great part of this knowledge is provided by vision. Similarly, making robots interact with their environment by using computer vision perception remains as one of the most exciting research subjects for the scientific community. Performing robotic tasks by using a comprehensive interpretation of the environment, as humans can do, is not yet realistic. Therefore, the tasks must be performed by using cues provided by lower levels of perception like colour, texture and shape recognition and depth perception. Many approaches derived from these types of visual perception have been applied in robotics during the last 20 years [DeSouza and Kak, 2002].

One of the approaches that have obtained more relevance in the last years is known as *visual servoing* or visual servo control [Hutchinson *et al.*, 1996]. This type of approach is based on executing robotic tasks by using a set of features provided by visual perception known as *visual features*. Concretely, the goal of the task, or desired state, is defined with the features as they are perceived in such state. Then, in any other state where the goal is not attained, the current perceived features are compared to the desired ones and an action suitable for converging to the desired state is generated. This process is made with a control loop based on visual feedback, in order to adapt to changes in the environment and to make more robust the execution of the task. In order to correctly generate the action, visual servoing models the link between the variation of the visual features and the robot kinematics. Typical applications of visual servoing are robot positioning with respect to static objects or target tracking.

One of the key points in visual servoing is the extraction of features from images, which is a classic problem of computer vision. For example, if points [Feddema *et al.*, 1991], lines [Andreff *et al.*, 2002] or regions [Chesi *et al.*, 2000] are used as features, it is necessary to extract the same elements in the image corresponding to the desired state and the image of the current state. In the beginning, the problem was usually simplified by positioning artificial landmarks in the scene. Nowadays more sophisticated computer vision algorithms have been adopted in visual servoing in order to deal with objects with complex textures or natural and unstructured scenes [Tahri and Chaumette, 2004; Collewet *et al.*, 2004]. Another key point in most visual servoing approaches is the need of matching the visual features between images taken in different states or different points of view. We remark that this problem has a great analogy with the correspondence problem related to passive stereovision. Therefore, visual servoing is unable to deal with non-textured objects, with objects for which extracting features is very complicated, or with adverse lighting conditions.

Coded structured light appears as a potential solution to the problems of visual servoing when dealing with non-textured or too complex objects. Furthermore, visual servoing based on features extracted from structured light has not been deeply studied up to date. Therefore, it seems very reasonably to study the potential contributions of a structured light approach to visual servoing. In this field, there are open issues that must be investigated like how structured light can be used for enlarging the application field of visual servoing and specially, how it can be used for optimising the link from sensing to acting in order to obtain a robust control law.

1.3   Context and motivations

This work has been developed in the context of several research projects from the Catalan, Spanish and French governments. The work has been funded by a fellowship from the Ministry of Universities, Research and Information Society of the Catalan Government.

A part of the work has been made within the VICOROB group[1] of the University

---

[1]Computer Vision and Robotics Group. vicorob.udg.es

of Girona, counting with a total of 26 members between researchers and PhD students. The research areas of the group are underwater robotics and vision, mobile robotics, $3D$ perception and image analysis. The research activities are currently supported by several national projects and an European project, like the development of autonomous underwater vehicles, monitoring the deep seafloor on the mid-Atlantic ridge or mammographic image analysis based on content. The work specifically developed in this work has been partially funded by the following Spanish projects:

- The MCYT[2] project TAP[3] 1999-0443-C05-01 from 31/12/99 until 31/12/02. The aim of this project was the design, implementation and accuracy evaluation of mobile robots fitted with distributed control, sensing and a communicating network. A computer vision based system was developed for providing the robots the ability of exploring an unknown environment and building a dynamic map. This project took part of a bigger project coordinated by the Polytechnic University of Valencia (UPV) and integrated both the Polytechnic University of Catalonia (UPC) and the University of Girona (UdG).

- The MCYT project TIC[4] 2003-08106-C02-02 from 01/12/03 until 30/11/06. The aim of the global project in which it is integrated is to design and develop FPGA-based applications with fault tolerance applied to active vision-based surveillance tasks in large surfaces like airports and train stations. Some of the tasks involved are automatic detection of dangerous situations or suspicious behaviours, and people tracking. The project is developed in collaboration with the UPV.

Some research areas in these projects were out of the field of the group. For example, the first project required to develop robot navigation tasks like obstacle avoidance based on computer vision while the second one requires to control a camera for surveillance applications. In both cases, active stereovision based on structured light was aimed to be integrated in order to enlarge the application field to unstructured environments. Furthermore, the requirements of these and other tasks seemed to fit on the field of visual servoing, which was not being studied in the VICOROB group. Then, the need of starting a collaboration with a first order research centre in this field arose. That was the origin of the collaboration with the LAGADIC group (initially integrated in the VISTA group) at the IRISA[5] in Rennes, in France. This group takes part of the French institute INRIA[6]. The main research axis of the LAGADIC group are visual servoing, image processing and $3D$ localisation applied to robotics, animation and augmented reality. The first collaboration consisted of a stay of 3 months between 2002 and 2003 where the fundamentals of visual servoing were studied under the supervision of Dr. François Chaumette, director of the group. At the end of the stay, a three-part collaboration between the VICOROB group,

---

[2]Ministerio de Ciencia y Tecnología
[3]Tecnologías Avanzadas de Producción
[4]Tecnologías de la Información y de las Comunicaciones
[5]Institut de Recherche en Informatique et Systèmes Aléatoires. www.irisa.fr
[6]Institut National de Recherche en Informatique et Automatique. www.inria.fr

the LAGADIC group and the CEMAGREF[7] of Rennes was planned for the following years.

The CEMAGREF is formed by several research centres in France where one of the research topics is the equipment engineering for the agrifood industry. PAIC [8] is one of the research groups in Rennes. One of its projects, directed by Dr. Christophe Collewet, is related to the traceability of meat products in an industrial chain by using vision and a robot manipulator. Due to the difficulty for positioning the robot with respect to products like pieces of pork, a solution based on combining structured lighting and visual servoing had been suggested. Then, due to the common research points between the problem being treated by the CEMAGREF and the research being developed at the VICOROB group, a natural collaboration arose thanks to the mediation made by the LAGADIC group.

A joint work was signed between the University of Girona and the University of Rennes. The aim of the work was directed to develop visual servoing approaches based on the projection of structured light for enlarging the field of robotic applications in unstructured or complex environments or adverse lighting conditions. A research project has been funded by the CEMAGREF from 01/01/04 until 31/12/05 in order to finance part of the research, technologic investments and conference attendance for dissemination of the scientific results. The aim of the project is to design, implement and integrate structured light sensors in a 6-degrees-of-freedom robotic cell for positioning tasks. Furthermore, another objective is to study the contribution of structured light for obtaining a robust control law in visual servoing.

Two times six months in Rennes were planned. The stays were supervised by Dr. Joaquim Salvi (from VICOROB), Dr. François Chaumette (from LAGADIC) and directed by Dr. Christophe Collewet (from CEMAGREF). The aim of these stays was to correctly develop the work by taking profit of the deep knowledge on structured light from the VICOROB group and the expertise on visual servoing of the LAGADIC and PAIC groups.

1.4 Objectives

The main objective of this work consists in developing techniques for visual control of robots based on the use of structured light. This goal is motivated by the need of executing positioning tasks in robotics with respect to non-textured objects or those for which extracting visual features is too complicated.

In visual servoing, the most typical configuration is based on attaching a camera to the end-effector of a robot. Then, the goal consists in "moving" the camera to a position where a given desired image is attained. When no visual features can be easily extracted from the images, the use of structured light can greatly simplify this problem.

In a first stage, it is necessary to study the large variety of patterns existing in structured light. The largest taxonomy of patterns appears in coded structured light. This evolution of classic structured light is able to obtain unambiguous correspondences by

---

[7]French Institute of Agricultural and Environmental Engineering Research. www.cemagref.fr
[8]standing for Perception, Image Analysis and Control

including a coding strategy on the projected patterns. A comprehensive knowledge on the state of the art of these techniques is fundamental in order to distinguish which patterns are suitable for being used in visual servoing applications.

The use of both structured light and coded structured light in a visual servoing framework needs further investigation. There are very few works addressing this combination. There is a lot of uncertainty on how these techniques can contribute to visual servoing. First of all, it seems clear that projecting light patterns can introduce visual features independently of the object appearance. A second more ambitious objective is to study whether a specific design of the projected pattern can lead to the optimisation of the visual based control law.

1.5 Book outline

At first a comprehensive study of coded structured light as a technique for solving the correspondence problem is addressed in Chapter 2. Coded structured light is a superset of techniques including the case of non-coded patterns like laser spots, planes, or grids. The inclusion of a coding strategy in the patterns increases the number of correspondences and removes ambiguities in their determination. A survey on the existing patterns is a key issue in order to distinguish which ones are the most suitable for being used in a visual servoing scheme and opens a new research area in this field. Another goal of this chapter is to evaluate the performance of every pattern in terms of resolution, i.e. the number of correspondences provided by the pattern, and their accuracy. This evaluation is made in a shape acquisition framework for $3D$ object reconstruction, which is the typical application of coded structured light. From the conclusions arising from the comparative results, we have realised that some patterns intended to reconstruct moving objects suffer of low resolution or they provide low accuracy in the measurements. For this reason, in Chapter 3 a new pattern is proposed, contributing to the field of coded structured light. The new pattern increases the resolution of the most usual type of one-shot techniques by obtaining good accuracy in the $3D$ reconstructed points. The new pattern is experimentally compared to similar existing patterns in order to validate the approach.

The use of visual servoing for robotics tasks by taking complex objects or objects lacking of visual features into account is either a complex problem or an impossible issue. The projection of coded light patterns on the scene for inserting visual features opens a new research area that must be studied. An application showing the advantages and drawbacks when using a coded light pattern for visually guiding a robot with classic visual servoing is presented in Chapter 4. The work focuses on an eye-in-hand configuration where a LCD projector is placed apart from the robot projecting a coded pattern on the working area. Furthermore, a discussion about the 3D reconstruction capabilities of a coded structured light setup in a visual servoing framework are also addressed.

The use of a deported structured light projector is suitable for certain applications, predominantly in industrial environments. However, it can be unsatisfactory for other applications when the whole working area cannot be covered by the field of view of the projector, or in mobile robotics when the working area is not a specific place. A possible solution to these cases consists in attaching the structured light emitter to the robot and the camera. Chapter 5 is devoted to design and implement a dedicated onboard structured light sensor for robot guidance which allows the execution of a specific positioning task. Furthermore, it is shown that a suitable design of the sensor leads to the optimisation of the control loop in terms of decoupling, stability and camera trajectory. This part of the work is supported by strong theoretical analysis and experimental results.

Finally, Chapter 6 presents the conclusions of the work, including a list of the related publications and conference contributions. Further work derived from results and some perspectives are also discussed.

Chapter 2

Solving the correspondence problem with coded patterns

Projecting structured light patterns onto the environment is a largely used technique in computer vision and robotics. The main advantage is that the projected visual features are easily distinguished by a camera. In order to avoid ambiguities and reliably solving the correspondence problem the patterns can be coded. Coded structured light is a technique based on projecting a light pattern and imaging the illuminated scene from one or more points of view. Since the pattern is coded, correspondences between image points and points of the projected pattern can be easily found. This chapter presents an overview of the existing techniques, as well as a new classification of patterns for structured light sensors. We have implemented a set of representative techniques in this field and here some comparative results are presented. The typical framework where coded light is used is in shape acquisition. Therefore, the techniques presented in this chapter are discussed and compared taking into account their 3D reconstruction performance which is strongly related to their performance on providing accurate correspondences.

2.1 Introduction to structured light

The term *structured light* is used to refer to a vision system taking profit of an active light source which projects a light pattern onto the environment. The pattern is intended to aid a computer vision task to be accomplished. The most typical application of structured light belongs to the active stereovision field for obtaining range measurements. In this case, the most typical configuration consists of a camera and a structured light emitter. Among all the ranging techniques [Jarvis, 1993; Chen *et al.*, 2000], stereovision is based on imaging the scene from two or more points of view and then finding pixel correspondences between the different images in order to triangulate the 3D position. Triangulation is possible if cameras are previously calibrated [Faugeras, 1993; Salvi *et al.*, 2002]. However, difficulties in finding the correspondences arise, even when taking into account epipolar constraints [Armangué and Salvi, 2003; Zhang, 1998]. The projection of structured light makes easier the correspondence problem either between multiple cameras or between a camera and the light source.

The source of light is typically a laser emitter with diffracting lenses allowing simple patterns to be projected like light spots, circles, lines and grids. Structured light can be both projected in the range of the visible and the invisible spectrum [Fofi *et al.*, 2004].

2.2 Typical application: shape acquisition

One of the fields of application of structured light is in robotics. In this case, structured light is used in range sensing [Nevado *et al.*, 2004; Matthies *et al.*, 2002; Kondo and Tamaki, 2004; Sun *et al.*, 2004; Sazbona *et al.*, 2005], SLAM[1] [De la Escalera *et al.*, 1996; Dubrawski and Siemiatkowska, 1998; Neira *et al.*, 1999; Kim and Cho, 2001; Surmann *et al.*, 2003; Jung *et al.*, 2004], obstacle avoidance [Le Moigne and Waxman, 1988; Weckesser *et al.*, 1995; Haverinen and Roning, 1998; Joung and Cho, 1998] and other robotic applications which will detailed in Chapter 4.

However, among all the applications of structured light the most typical one is $3D$ reconstruction obtained by triangulation. Therefore, structured light is often used in a shape acquisition system which is a topic with a large variety of applications. Some of them are dense range sensing, industrial inspection, object recognition, $3D$ map building, reverse engineering, fast prototyping and even preservation of cultural heritage [Levoy *et al.*, 2000] and animation [Zhang *et al.*, 2004].

The main advantage of structured light in front of passive stereovision is that the search of correspondences is greatly simplified since it must be done only for image regions where the reflected pattern appears. Furthermore, structured light allows correspondences on non-textured objects to be found. The first shape acquisition systems based on structured light were laser scanners [Forest and Salvi, 2002]. These devices are typically based on scanning the object with a laser plane and detecting the projected line in the camera image for triangulating all the illuminated points. The advantage of these scanners is the large resolution and accuracy obtained leading to high quality $3D$ surface reconstruction. The main drawback is that they are limited to static objects and that a large number of images must be acquired. Furthermore, in order to scan the object either the laser plane must be rotated, or both the laser and the camera or the object must be moved at each iteration. In the latter case, the displacement must be known so that free-moving objects cannot be reconstructed. All these problems appear because in each acquired image only few points can be triangulated, i.e the points belonging to the laser stripe. This limitation can be minimised by projecting more complex patterns like a laser grid [Le Moigne and Waxman, 1988]. However, a new problem arises: since the grid has a unique colour the identification of every grid region in the image becomes ambiguous.

In order to solve all these limitations coded structured light appeared as a flexible alternative [Batlle *et al.*, 1998; Salvi *et al.*, 2004]. Coded structured light consists in using

---

[1]Simultaneous Localisation and Mapping

a device that projects a light pattern onto the measuring surface (see Figure 2.1) instead of using lasers. The most commonly used devices are LCD video-projectors, although previously the most typical were slide projectors. Such devices project an image with a certain structure so that a set of pixels are easily distinguishable by means of a local coding strategy. Thus, locating such coded points in the camera image is solved directly with no need for geometrical constraints. The projecting images are called patterns, since they present a globally structured appearance. The simplest pattern is a black image with an illuminated pixel. In this case, only one point can be reconstructed by triangulation by using the pixel coordinates of the illuminated point in the pattern and the corresponding coordinates in the camera image. Note that this case is equivalent to use a camera and a laser pointer. In general, all the patterns available with laser technology can be reproduced with a video-projector.

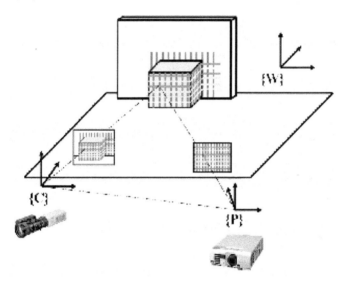

Figure 2.1: Schema of a coded structured light system.

This chapter presents a comprehensive survey on coded structured light techniques and proposes a new consistent classification. The chapter focuses on the different coding strategies used in the bibliography and reproduces the experimental results of several techniques in order to evaluate and compare their accuracy and analyse their applicability. The comparison framework is a shape acquisition setup.

The chapter is structured as follows: firstly, the classification is presented in section 2.3. Secondly, in section 2.4 techniques based on projecting multiple patterns are explained. In section 2.5, techniques exploiting the spatial neighbourhood paradigm are presented. Next, in section 2.6, coding strategies using direct codification are also explained. In section 2.7, the experimental results obtained with a set of implemented techniques are presented. Concluding, in section 2.8, a discussion about the advantages and drawbacks of every subgroup of techniques is included. In addition, general guidelines for choosing the most suitable technique, given the specifications of an application, are proposed.

2.3 A classification of coding strategies for structured light patterns

A coded structured light system is based on the projection of a single pattern or a set of patterns onto the measuring scene which is then viewed by a single camera or a set of cameras. The patterns are specially designed so that codewords are assigned to a set of pixels. Every coded pixel has its own codeword, so there is a direct mapping from the codewords to the corresponding coordinates of the pixel in the pattern. The codewords are simply numbers, which are mapped in the pattern by using grey levels, colour or geometrical representations. The larger the number of points that must be coded, the larger the codewords are and, therefore, the mapping of such codewords to a pattern is more difficult. The aim of this work is to review the available strategies used to represent such codewords.

Pattern projection techniques differ in the way in which every point in the pattern is identified, i.e. what kind of codeword is used, and whether it encodes a single axis or two spatial axis. In reality, it is only necessary to encode a single axis, since a $3D$ point can be obtained by intersecting two lines (i.e. when both pattern axis are coded) or intersecting one line (the one which contains a pixel of the camera image) with a plane (i.e. when a single pattern axis is coded).

Table 1 shows pattern projection techniques classified according to their coding strategy: time-multiplexing, neighbourhood codification and direct codification. The seven columns on the right of the table indicate whether or not a given pattern is suitable for measuring moving objects, the colour depth used and whether repeated codewords appear (periodic codification) or not (absolute codification). Time-multiplexing techniques generate the codewords by projecting a sequence of patterns along time, so the structure of every pattern can be very simple. Furthermore, in spite of increasing the pattern complexity, neighbourhood codification represents the codewords in a unique pattern. Finally, direct codification techniques define a codeword for every pixel, which is equal to its grey level or colour.

In the following sections, each one of these three classifying groups are explained in detail. Moreover, the different techniques proposed in the bibliography which belong to each subgroup are introduced, and we will show the evolution from the simplest to the most complex technique.

2.4 Time-multiplexing strategy

One of the most commonly used strategies is based on temporal coding. In this case, a set of patterns are successively projected onto the measuring surface. The codeword

Table 2.1: The proposed classification of coding strategies.

Strategy	Method	Author	Static	Moving	Binary	Grey levels	Colour	Periodical	Absolute
Time-multiplexing	Binary codes	Posdamer et al.	√		√				√
		Inokuchi et al.	√		√				√
		Minou et al.	√		√				√
		Trobina	√		√				√
		Valkenburg and McIvor	√		√				√
		Skocaj and Leonardis	√		√				√
		Rocchini et al.	√				√		√
	n-ary codes	Caspi et al.	√				√		√
		Horn and Kiryati	√			√			√
		Osawa et al.	√			√			√
	Gray code + Phase shifting	Bergmann	√			√			√
		Sansoni et al.	√		√				√
		Wiora	√			√			√
		Gühring	√			√			√
	Hybrid methods	Kosuke Sato	√		√	√			√
		Hall-Holt and Rusinkiewicz		√	√				√
		Wang et al.		√	√				√
		Guan et al.		√		√			√
Spatial Neighborhood	Non-formal codification	Maruyama and Abe		√	√			√	
		Durdle et al.		√		√		√	
		Ito and Ishii		√		√			√
		Boyer and Kak		√			√		√
		Chen et al.		√			√		√
		Koninckx et al.		√		√	√		√
	De Bruijn sequences	Hügli and Maître		√			√		√
		Monks et al.		√			√		√
		Vuylsteke and Oosterlinck		√	√				√
		Salvi et al.		√			√		√
		Lavoie et al.		√			√		√
		Zhang et al.	√	√			√		√
	M-arrays	Morita et al.		√		√			√
		Petriu et al.		√	√				√
		Kiyasu et al.		√	√				√
		Spoelder et al.		√	√				√
		Griffin and Yee		√	√		√		√
		Davies and Nixon		√			√		√
		Morano et al.	√		√		√		√
Direct coding	Grey levels	Carrihill and Hummel	√			√			√
		Chazan and Kiryati	√			√			√
		Hung		√		√		√	
	Colour	Tajima and Iwakawa	√				√		√
		Smutny and Pajdla	√				√		√
		Geng		√			√		√
		Wust and Capson		√			√	√	
		Tatsuo Sato	√				√	√	

Scene applicability	Static	
	Moving	
Pixel depth	Binary	
	Grey levels	
	Colour	
Coding strategy	Periodical	
	Absolute	

for a given pixel is usually formed by the sequence of illumination values for that pixel across the projected patterns. Therefore, the codification is called *temporal* because the bits of the codewords are multiplexed in time. This kind of pattern can achieve high accuracy in the measurements. This is due to two factors: first, as multiple patterns are projected, the codeword basis tends to be small (usually binary) and therefore a small set of primitives is used, which are easily distinguishable among each other; secondly, a coarse-to-fine paradigm is followed, as the position of a pixel is encoded more precisely while the patterns are successively projected.

During the last twenty years several techniques based on time-multiplexing have appeared. We have classified these techniques as follows: a) techniques based on binary codes: a sequence of binary patterns is used in order to generate binary codewords; b) techniques based on n-ary codes: a basis of n primitives is used to generate the codewords; c) Gray code combined with phase shifting: the same pattern is projected several times, shifting it in a certain direction in order to increase resolution; d) hybrid techniques: a combination of time-multiplexing and neighbourhood strategies.

The following sections describe in detail the techniques which can be included in such coding strategies.

2.4.1 Techniques based on binary codes

In these techniques only two illumination levels are commonly used, which are coded as 0 and 1. Every pixel of the pattern has its own codeword formed by the sequence of 0s and 1s corresponding to its value in every projected pattern. Therefore, a codeword is obtained only when the sequence is completed. An important characteristic of this technique is that only one of the two pattern axis is encoded.

Posdamer and Altschuler [Posdamer and Altschuler, 1982] were the first to propose the projection of a sequence of m patterns to encode 2^m stripes using a plain binary code. Therefore, the codeword associated with each pixel is the sequence of 0s and 1s obtained from the m patterns, the first pattern being the one which contains the most significant bit. In this case, 16 columns of the projector image are coded. The symbol 0 corresponds to black intensity while 1 corresponds to full illuminated white. Therefore, the number of stripes increases by a factor of two for each consecutive pattern. Every stripe of the last pattern has its own binary codeword. The maximum number of patterns that can be projected is the resolution in pixels of the projector device. However, reaching this value is not recommended because the camera cannot always perceive such narrow stripes. It should be noted that all pixels belonging to the same stripe in the highest frequency pattern share the same codeword. Therefore, it is necessary to calculate either the centre of every stripe or the edge between two consecutive stripes. The latter has been shown to be the best choice. In Figure 2.2a a sequence of 4 patterns binary encoded are shown.

Inokuchi et al. [Inokuchi *et al.*, 1984] improved the codification scheme of Posdamer and Altschuler by introducing Gray code instead of plain binary. The advantages of Gray code is that consecutive codewords have a Hamming distance of one, being more robust against noise. In Figure 2.2b, the corresponding Gray coded patterns can be observed.

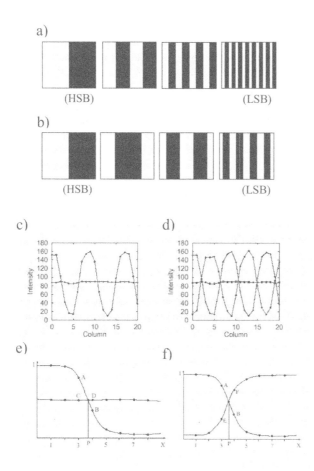

Figure 2.2: Stripe patterns coded with binary codes and stripe edges detection: a) Patterns coded with plain binary; b) Gray code patterns; c) Variant binary threshold of normal stripe pattern; d) Variant binary threshold of both normal and inverse stripe patterns; e) Stripe position by normal stripe pattern and binary threshold; f) Stripe position using normal and inverse patterns.

Minou et al. [Minou *et al.*, 1981] designed another technique based on time coded parallel stripes. The aim was to create a depth measurement system which was robust in the presence of noise. For this purpose, the authors decided to use both binary code and the Hamming error correcting code. The number of coded stripes was only 25, therefore, the plain binary code had length 5 and the correcting code had length 9. It should be noted that the number of coded stripes is very small due to the large amount of bits needed to create a code with a Hamming distance of three, which then allows a single error correction.

Trobina presented an error model of coded light range sensors based on Gray coded patterns [Trobina, 1995]. The author demonstrated that the crucial step of these sensors is the accurate location of every stripe in the image. In the finest pattern only half of all the edges can be measured, while the other half can be found in the previous patterns. By simple bifurcation of the images the stripes can be found. The binary threshold is fixed for every pixel independently. It is necessary to acquire images of fully illuminated scenes (white pattern) and non-illuminated scenes (black pattern). The variant threshold is the mean between the grey level of such images as shown in Figure 2.2c. Hence, with such bifurcation, the edges can be detected with *pixel accuracy*. However, the profile of the transition between a white and a black stripe in the images is not a perfect step. It is normally a non linear profile. Two ways of detecting the edges with *sub-pixel accuracy* were proposed.

The first way of detecting stripe edges with sub-pixel accuracy is to find the zero-crossings of the second derivative of the image, orthogonally to the stripes. The problem with this approach is finding the optimal gradient filter size. An alternative way is to project both normal and inverse stripe patterns, i.e. positive and negative patterns. Then, by finding the intersection of both profiles, the stripe edge is located. Since the profiles are non-linear functions, linear interpolation is used among the nearest sample points (grey levels of nearby pixels). As shown in Figure 2.2f, by intersecting line AB with line EF, the edge is located. As can be seen in Figure 2.2d, the intersection of both inverse and normal profiles do not always coincide with the variant binary threshold, so this method is more accurate. If projecting inverse patterns is not desired, the linearly interpolated normal profile can be intersected with the variant threshold profile. This technique is shown in Figure 2.2e, where the segment AB should be intersected with segment CD. After experimental results, Trobina concluded that linear interpolation is more accurate than 2nd derivative and the best results are obtained if both normal and inverse patterns are projected.

Locating the stripes accurately when projecting Gray coded patterns was also the main objective of the work presented by Valkenburg et al. [Valkenburg and McIvor, 1998]. In this case, every acquired image is divided into regions of 17×17 pixels. In every region, a $2D$ third order polynomial is interpolated by means of least square fitting, obtaining a facet that approximates all the stripes in the region of interest. The authors also made experiments fitting sinusoidal functions in the regions, slightly improving the results.

Objects containing regions with different reflective properties are difficult to reconstruct. When projecting patterns at low illumination intensities, the signal-to-noise ratio of the system decreases and, therefore, depth from low reflective regions cannot be obtained. On the other hand, when projecting high illumination intensity patterns, depth from regions with high reflectance cannot be recovered due to pixel saturation. So, most binary coded techniques assume that the objects have uniform albedo, otherwise, the whole surface cannot be reconstructed. Skocaj and Leonardis [Skocaj and Leonardis, 2000] proposed a new strategy to overcome these limitations by increasing the number of projected patterns. Projecting multiple images at different illumination intensities of a given stripe pattern allows each view of such patterns to be combined into a single *radiance map*. A radiance map contains for, each pixel, the relative reflective factor of the corresponding surface points. In general, we have that

$$g \propto r \cdot l \tag{2.1}$$

where g is the pixel value, r is the reflective of the corresponding surface point and l is the illumination intensity incident to the surface point. If relative radiance values are considered, then for every pixel i in the image j the relation expressed in eq. 2.2 exists.

$$r_{ij} = g_i/l_j \tag{2.2}$$

The variation of the illumination in the scene should not affect the reflective value of the surface point. However, the relationship is non-linear due to the distortions introduced by both the projector and the camera. In order to eliminate this non-linearity, an equation system is defined which considers all the pixels under different illumination values. The overdetermined system is minimised by using least-squares. Then, the best fitting reflectance value of every corresponding surface point is obtained. Hereafter, a global radiance map can be defined containing the reflectance values relating to every pixel of the image. Using this radiance map, the projected intensity l_j can be inversely recovered for every pixel value g_{ij} and its associate reflectance value r_i. The minimum number of illumination intensities to be projected is two (binary). However, by projecting more intensities, better results are obtained. In order to calculate the range image, the sub-pixel localisation of the stripe edges proposed by Trobina [Trobina, 1995] was applied. The contribution of this work was a simultaneous reconstruction of both high and low reflective surfaces.

During the last few years, most of the work dealing with binary coded patterns has been aimed at improving the sub-pixel localisation of stripe edges. Rocchini et al. [Rocchini *et al.*, 2001], introduced a slight change in the typical Gray coded patterns in order to ease the localisation of stripe transitions. For this purpose, they proposed encoding the stripes with blue and red instead of black and white. Moreover, a green slit of a pixel width was introduced between every stripe. Then, the stripe transitions of the higher resolution pattern are reconstructed by locating the green slit with sub-pixel accuracy.

2.4.2   Techniques based on n-ary codes

The main drawback of the schemes based on binary codes is the large number of patterns which need to be projected. However, the fact that only two intensities are projected eases the segmentation of the viewed patterns. There are some works which consider the problem of reducing the number of patterns by means of increasing the number of intensity levels used to encode the stripes. There now follows an explanation of the two formal schemes of increasing the coding basis.

Caspi et al. [Caspi *et al.*, 1998] proposed a multilevel Gray code based on colour. The extension of the Gray code is based on an alphabet of n symbols, where every symbol is associated to a certain RGB colour. This extended alphabet makes it possible to reduce the number of patterns. For example, with binary Gray code, m patterns are necessary to encode 2^m stripes. With an n-Gray code, n^m stripes can be coded with the same number of patterns.

This work was very important since it is the generalisation of the most widely used coding strategy in the time-multiplexing paradigm. The n-ary code shares the same characteristics of a binary Gray code by fixing a Hamming distance of one between consecutive codewords. The work by Caspi et al. not only develops the mathematical basis for generating n-ary codes, but also analyses the illumination model of a structured light system. This model takes into account the light spectrum of the LCD projector, the spectral response of a 3-CCD camera and the surface reflectance. The whole model is presented in eq. 2.3.

$$
\underbrace{\begin{bmatrix} R \\ G \\ B \end{bmatrix}}_{\vec{C}} = \underbrace{\begin{bmatrix} a_{rr} & a_{rg} & a_{rb} \\ a_{gr} & a_{gg} & a_{gb} \\ a_{br} & a_{bg} & a_{bb} \end{bmatrix}}_{A} \underbrace{\begin{bmatrix} k_r & 0 & 0 \\ 0 & k_g & 0 \\ 0 & 0 & k_b \end{bmatrix}}_{K} \vec{P} \underbrace{\left\{ \begin{matrix} r \\ g \\ b \end{matrix} \right\}}_{\vec{c}} + \underbrace{\begin{bmatrix} R_0 \\ G_0 \\ B_0 \end{bmatrix}}_{\vec{C}_0} \tag{2.3}
$$

where $\vec{c}$ is the projection instruction for a given stripe, i.e. $[0\ 0\ 0]^T$ represents black and $[255\ 255\ 255]^T$ is white; $\vec{P}$ is the non-linear transformation from projection instruction to actually projected intensities for every RGB channel; and A is the projector-camera coupling matrix. Every element a_{ij} of matrix A is the convolution of the camera spectral response for channel i with the spectrum of the light projected in channel j. Matrix A shows the crosstalk between colour channels. $\vec{C}$ is the vector containing the RGB camera readings of a certain pixel. $\vec{C}_0$ is the camera readings corresponding to the scene under ambient lighting. Finally, K is the surface reflectance matrix specific to every scene point projected into a camera pixel. This matrix contains a reflectance constant for every RGB channel.

The main benefit of this model is that it considers a constant reflectance for every scene point in the three RGB channels. This is much more realistic than considering colour neutrality of the scene, which is commonly assumed in most systems dealing with colour coding schemes. In the case of colour neutrality, matrix K is the identity for every pixel.

In order to calculate the different terms in eq. 2.3 it is necessary to fulfill a colourimetric calibration. With this procedure, A and $\vec{P}$ are obtained. $\vec{P}$ is a non-linear function, but invertible, so it can be implemented in three look up tables (one for every colour channel). The colourimetric calibration is only necessary once. Then, matrix K is obtained by just taking a reference image under white illumination ($\vec{c_w} = [255\ 255\ 255]^T$) and solving eq. 2.3. Some approximations of the model can be done in order to avoid the whole colourimetric procedure, as explained in [Zhang *et al.*, 2002].

The illumination model proposed allows the projected colour to be estimated from camera readings. This is very important when working with colour encoded stripes, since correct identification of colours leads to correct codewords. Therefore, such a model can be applied to any system dealing with colour.

Another important aspect of Caspi's work is its adaptation to the environment. This means that the system can be configured with different parameters. In this case, the parameters are the number of patterns to be projected M, the number of colours used L, and the noise immunity factor α. If M and L are chosen, then α is fixed. Otherwise, if α and L are chosen, then M is fixed. The noise immunity factor α determines the distance between adjacent consecutive codewords in every RGB channel. The higher α is, the more robust is the colour identification and the higher the number of patterns M since less colours can be used or fewer stripes need to be coded. After experiments, Caspi et al. determined that the n-ary Gray codes achieve the accuracy and robustness of the binary Gray code technique using fewer patterns.

Another technique that encodes adjacent stripes with n-ary codewords is the one proposed by Horn and Kiryati [Horn and Kiryati, 1999]. The alphabet of the codes is based on multiple grey levels instead of a binary alphabet. The aim of the work was to find the smallest set of patterns that meet the accuracy requirements of a certain application producing the best performance under certain noise conditions.

Given an alphabet of n symbols, a code is created so that consecutive codewords have a Hamming distance of one. Every element of the alphabet is mapped to a certain grey level. When expressing the differing elements of two consecutive codewords in terms of the associated grey levels, the difference is constant for all pairs of consecutive codewords. For example, if 256 grey levels are available, when using a binary Gray code the distance of consecutive codewords in terms of the grey levels is 255 or 100%. By increasing the basis of the code and maintaining the length, more codewords can be generated in spite of decreasing the distance in terms of % of available grey levels between consecutive ones. The authors proposed the use of space filling curves such as Hillbert or Peano curves [Sagan, 1994] for defining the code. Such curves represent a path in an n-dimensional space, passing through a set of points so that consecutive points are joined by straight segments. The distance between consecutive points is constant along all the curve. Like the system proposed by Caspi et al., there are several parameters which can be tuned, so that L is the number of stripes to encode; K the number of projecting patterns, and therefore the length of the codewords; m the order of the curve used to place the codewords; and r the desired distance between consecutive codewords (% of the total of grey levels), which is proportional to the noise immunity factor of the system. The K parameter is also the dimension of the space filling curve that will be used to generate the code. If parameters K and m are fixed, then the larger the parameter r is, the more robust the resulting codification is against noise, but the number of stripes, L, decreases. Therefore, there is a trade off between noise immunity, number of patterns, distance between codewords and number of encoded stripes. For a given r, if the number of stripes L must be increased it means that the length of the space filling curve must also grow. To achieve this, there are two possible solutions: increasing the curve order, which has the problem of reducing the distance between consecutive codewords (since the distance between consecutive points of the curve also reduces); or increasing the dimension of the curve, i.e. increasing the number of projecting patterns.

In Figure 2.3a, a $3D$ Hillbert curve of the 2nd order is shown. Every dimension of the curve is associated with one of the patterns to be projected. The number of stripes to encode has been fixed at $L = 128$, so a total number of 128 3D points have been placed equidistantly along the curve. Consecutive points along the curve correspond to adjacent stripes in the patterns. The value of every point component is the grey level associated with the stripe in one of the patterns. Therefore, every point in the curve produces the codeword of grey levels for the corresponding stripe. The number of grey levels used in the example is 7. The extracted intensity profiles of every three patterns are shown in Figure 2.3b, while the resulting patterns are shown in Figure 2.3c. Every intensity profile is the projection of the points in the curve in one of the three axis: $f1(x)$, $f2(x)$ and $f3(x)$ are the intensity profiles of patterns 1, 2 and 3, respectively.

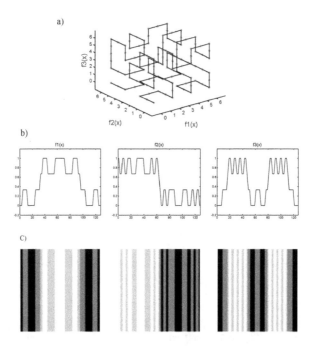

Figure 2.3: Horn and Kiryati coding strategy: a) 3-D 2nd order Hillbert curve with 128 codewords placed on it; b) Intensity profiles of the patterns extracted from the Hillbert curve; c) The resulting patterns.

Horn and Kiryati tested their system with a $3D$ Hillbert 2nd order curve where 256 codewords were placed. Therefore, the number of patterns was 3 and the number of grey levels 13. Such configuration produced better performances than a binary Gray coding scheme based on the projection of 9 patterns (512 stripes encoded). The design of coded patterns proposed by Horn and Kiryati produces more accurate results and reduces the number of projecting patterns.

2.4.3   Combination of Gray code and Phase shifting

Patterns based on Gray code, as well as binary and n-ary codes, have the advantage that the pixel codification is made punctually and no spatial neighbourhood has to be considered in the codification. However, the discrete nature of such patterns limits the range resolution. Furthermore, phase shifting methods exploit higher spatial resolution as they project a periodic intensity pattern several times by shifting it in every projection. The drawback of these methods is the periodic nature of the patterns, which introduces ambiguity in the determination of the signal periods in the camera images. The integration of Gray Code Methods (GCM) with Phase Shift Methods (PSM) brings together the advantages of both strategies, i.e. the unambiguity and robust codification of pattern stripes of GCM, plus the high resolution of PSM. The combination of both techniques leads to highly accurate $3D$ reconstruction. However, the number of projecting patterns increases considerably. We will now discuss some of the coded structured light systems that use this approach.

In [Bergmann, 1995], Bergmann designed a technique where some Gray coded patterns are projected in order to label the measuring surface regions where every period of a sinusoidal intensity pattern will be projected. Therefore, the ambiguity problem between signal periods is resolved. The sinusoidal patterns are represented by grey levels. A total number of four Gray patterns are projected in order to label 16 different regions on the measuring surface.Then, the periodic intensity pattern is projected four times by shifting $1/4$ of the period, each time.For every given pixel (x,y) of the camera image, the phase of the first periodic pattern projected to the corresponding surface point must be found. For this purpose, a classic four-step phase shift is applied in eq. 2.4. I_1, I_2, I_3 and I_4 are the grey levels of pixel (x,y) from camera images corresponding to every one of the 4 projected shifted patterns. Once the phase of a given pixel is known, the period of the sinus where the pixel lies is obtained with the Gray code labeling. Therefore, the pattern stripe projected to a certain surface point can be precisely calculated.

$$\phi(x,y) = \arctan\left(\frac{I_2 - I_4}{I_3 - I_1}\right) \tag{2.4}$$

Sansoni et al. compared the accuracy of GCM and PSM separately [Sansoni et al., 1997]. After experiments, they realised that both PSM and GCM obtain similar precision in their measurements (about 0.18mm). However, the resolution of PSM showed to be about 0.01mm in front of the 0.22mm of GCM. However, PSM failed in the steep slope changes at the measuring surface borders due to the occlusion of some periods of the PSM patterns. One interesting feature of the PSM patterns used by Sansoni et al. is that they are discrete stripe patterns with rectangular profiles. The sinusoidal profile is achieved by defocusing the LCD projector. When combining both methods, the mean error of the measurements was about $40\mu m$ with a standard deviation of $\pm 35\mu m$. Furthermore, Georg Wiora discussed the suitability of using LCD projectors when applying PSM alone or combined with GCM [Wiora, 2000]. He argued that such devices do not have enough contrast and radiant flux for PSM patterns with large resolution. According to the author, the best devices to use are special slide projectors, which allow 26000 black and white stripes to be projected on a 13mm slide (for more details refer to the article). Moreover, Wiora's article discusses the problems of mechanical misalignment of slides for these devices as well as problems of non-sinusoidal phase shift patterns.

Gühring proposed substituting the PSM for a new method called *line shifting* [Gühring, 2001], which was also combined with GCM. Gühring pointed out that *Phase Shifting* has a series of drawbacks. For example, when reconstructing surfaces with non-uniform albedo (with sharp changes from black to white) the phase cannot be determined precisely. Moreover, camera devices tend to integrate over a certain area so that pixel values are affected by its neighbours. To avoid this problem, camera resolution must be sufficiently higher than projector resolution. In order to avoid the problems of *Phase Shifting*, the author proposed substituting the sinusoidal periodic profile of such methods by a *multistripe* pattern, shifted several times. Gühring designed a 640×640 pattern for LCD projectors, where every 6th column is white and the remaining ones are black. By consecutively shifting the pattern 6 times, the whole resolution for every row of the pattern is covered. While repeating the process with row-encoded patterns, the entire resolution of the pattern is used. Since a multistripe pattern is also a periodic pattern (of discrete nature), the projection of Gray code patterns are also required in order to solve the ambiguities that arise.

To summarise, 32 patterns were projected onto the measuring surface: 9 vertical Gray codes; 6 vertical multistripe patterns, each one shifted a column towards the right; 9 horizontal Gray coded patterns; 6 horizontal multistripe patterns each one shifted one row downwards; 2 additional patterns for grey level normalisation (one fully illuminated and the other with the lamp switched off). With regard to the Gray coded patterns, a total number of n regions should be labelled, being n the number of lines projected in every *line shifting* pattern. Therefore, every pattern area where a line is shifted has its own label. For example, if patterns of 32 columns were projected, 6 lines would be shifted and, therefore, 6 bands of 6 pixels width should be labelled by the Gray code. Therefore, three patterns of Gray code should be projected. However, in the transitions of each region, a decoding error implies a large measuring error of around one period. That is why Gühring decided to introduce an oversampling technique consisting in projecting an additional Gray coded pattern. In this way, thinner bands of pixels are labelled, and more robust decodification of the regions is obtained. The maximum error when calculating the global position of an illuminated line due to transition of a Gray codeword was 2 pixels. Without the redundancy introduced by this oversampling, the same error rose to 6 pixels. In Figure 2.4a, the case of a row of a 32-pixel-wide pattern is shown, with the four patterns of Gray code and the 6 patterns containing line shifting. As can be seen, every region where a line is shifted contains three Gray codewords.

As for the patterns containing *line shifting*, the peak position of every illuminated line was intended to be located in the camera images with sub-pixel accuracy. Since the intensity profile of such perceived lines presented a gaussian distribution, the peak detector proposed by Blais and Rious [Trucco *et al.*, 1998] was used. This detector applies a linear derivative filter (higher order filters can be applied) at each pixel of every image row (when treating vertical multistripe patterns). The result obtained for each row is a set of local maxima and minima indicating the transitions from black to white regions and vice versa. Afterwards, for each pair of consecutive maximum and minimum, the zerocrossing of the linear interpolation between them is calculated and the sub-pixel position of the intensity peak is obtained. The detection process of a peak for a certain image row is shown in Figure 2.4b.

Gühring's line shifting method had a similar or even better resolution than techniques based on PSM and more accurate measurements. Note that this approach was inspired by traditional laser scanner techniques, which have been shown as the most accurate $3D$ profilers. The author developed a system set up based both on LCD and DMD projectors, obtaining similar results with an average error of $30\mu m$ and a maximum deviation of $0.281mm$ for both devices.

2.4.4   Hybrid methods

In the bibliography, there are some methods which are based on multiple pattern projection, so they use time-multiplexing, but also take into account spatial neighbourhood information in the decoding process. For example, the idea of Kosuke Sato [Sato, 1996] consisted of designing a certain binary pattern whose rows have a sharp impulse on its auto-correlation function. The pattern is projected several times by shifting it horizontally several times (the more times the pattern is shifted, the greater the resolution obtained). For every projection, an image is grabbed, in which the maximum autocorrelation peak of every row is computed. Then, as the phase shift of the corresponding projected pattern is known, the pixels containing such peaks can be reconstructed by triangulation. According to the author, this strategy achieves better accuracy than projecting and shifting a single slit, since the peak of cross-correlation shows sharper impulse and can be located more precisely.

Hall-Holt and Rusinkiewicz [Hall-Holt and Rusinkiewicz, 2001] divided four patterns into a total of 111 vertical stripes that were painted in white and black. Codification is located at the boundaries of each pair of stripes. The codeword of each boundary is formed by 8 bits. Every pattern gives 2 of these bits, representing the value of the bounding stripes. The most interesting aspect about this method is that it supports smooth moving scenes, something unusual in the time-multiplexing paradigm. This capability is due to a stage of boundary tracking along the patterns.

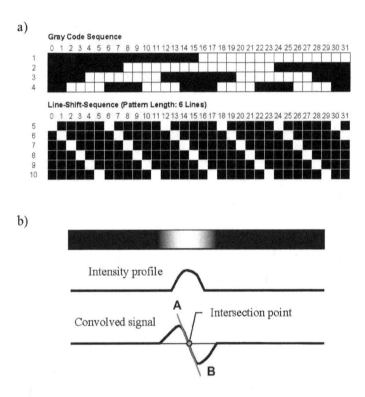

Figure 2.4: Line shifting technique: a) Gray code and line shifting patterns for 32-pixel-wide patterns; b) Intensity profile of a camera illuminated line and the peak detection with sub-pixel accuracy.

2.5 Spatial neighbourhood

The techniques in this group tend to concentrate all the coding scheme in a unique pattern. The codeword that labels a certain point of the pattern is obtained from a neighbourhood of the points around it. However, the decoding stage becomes more difficult as the spatial neighbourhood cannot always be identified and false correspondences can arise. Normally, the visual features gathered in a neighbourhood are the intensity or colour of the pixels or groups of adjacent pixels around it.

These spatial neighbourhood techniques can be classified as follows: a) strategies based on non-formal codification: the neighbourhoods are generated intuitively; b) strategies based on De Bruijn sequences: the neighbourhoods are defined using pseudorandom sequences; or c) strategies based on M-arrays: extension of the pseudorandom theory to the 2-D case.

In the following subsections some techniques from these three subgroups are summarised.

2.5.1 Strategies based on non-formal codification

Some authors have proposed techniques based on patterns designed so that it is divided into a certain number of regions, in which some information generates a different codeword, without using any mathematical coding theory. For instance, Maruyama and Abe [Maruyama and Abe, 1993] designed a binary pattern coded with vertical slits containing randomly distributed cuts, see Figure 2.5. The system was designed for measuring surfaces with smooth depth changes. The random cuts generate a set of linear segments so that the position of a segment in the pattern is determined by its own length and the lengths of the 6 adjacent segments. The decoding stage starts by matching every segment of the pattern with the observed slits of equal length. Multiple matchings can be found for every segment. In order to find out the correct matching, the lengths of the 6 adjacent segments must be considered. Once all the perfect matchings have been found, a region growing algorithm is applied in order to identify unmatched segments. The main drawback of this technique is that segment lengths can vary depending on the distance between the camera and the surface and the optics of both the camera and projector. All these factors considerably limit the robustness and the reliability of the system.

Some years later, a periodic pattern composed of the horizontal slits encoded with three grey levels was proposed by Durdle et al. [Durdle *et al.*, 1998]. The pattern is formed by the sequence $BWGWBGWGBGWBGBWBGW$, where B is a band of 4 black pixels, W is a band of 4 white pixels and G is a band of 4 half bright pixels. This sequence is repeated in the pattern until it covers all the vertical resolution. Due to the periodicity of the pattern, discontinuities larger than the pattern period are not permitted. The decoding stage of the system is composed of two steps: firstly, the starting point of every pattern period is searched in the grabbed image for every column, by finding the correlation peaks between the image column and a template of the projected period. Secondly, for every period, another template matching is made in order to find the subsequences WBG, WBG and

GWB. Repeating these processes for every image column, a large set of correspondences is found.

Ito and Ishii presented a *three-level checkerboard* pattern [Ito and Ishii, 1995]. The proposed pattern is a unique grid where each square is painted with one out of three possible intensity levels. The intensity level of a cell is chosen so that it is different from its four immediate neighbours. A *node* is defined as an intersection between four cells of the checkerboard. The *main code* of a node is identified by the intensity levels of its four adjacent cells. Since three intensity levels are used, the number of different main codes is 18 ($3 \times 2 \times 2 \times 2$). The *subcode* of a node is defined as the clockwise combination of the main codes of the four adjacent nodes. The subcode constitutes the codeword for every edge intersection of the pattern. Therefore, in order to decode the position of an observed node, it is necessary to analyse a spatial neighbourhood of 12 cells of the grid. Note also that both spatial coordinates of the nodes are encoded. In order to get a robust coding scheme, every possible subcode should appear only once in the pattern. However, Ito and Ishii were not studying the automatic generation of such a pattern, so they decided to allow repetitions of subcodes. The authors used epipolar restrictions between the camera and the projector in order to differentiate between nodes which share the same subcode.

The technique proposed by Boyer and Kak [Boyer and Kak, 1987] uses a pattern formed by vertical slits coded with the three basic colours (red, green and blue) and separated by black bands. The sequence of coloured slits was designed so that if the pattern is divided into subpatterns of a certain length, none is repeated. The most interesting thing about this work is the decoding stage. Boyer and Kak realised that the morphology of the measuring surface acts as a perturbation applied to the projected pattern (which acts like a signal), so the received pattern can contain disorders or even deletions of the slits. In order to match each received slit with the corresponding projected slit, a four-step algorithm was designed called *stripe indexing process*.

The first step is called *correlation*. Each unique subpattern of the original pattern is sliced along the received pattern to find all the positions where a perfect match takes place. Secondly, a region growing process of the matched subpatterns is carried out and tries to cover as many correspondences of slits as possible. This subprocess was called *crystal growing*. Thirdly, a fitting process is applied in order to remove erroneous matchings. When two subpatterns overlap, the thinnest is cut so that the shared slits are only

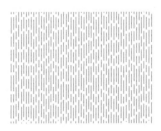

Figure 2.5: Pattern designed by Maruyama and Abe.

associated with the largest one. Finally, the matched slits are indexed. The whole process must be done for every epipolar line of the camera image. The authors did not take into account the information obtained from the previous epipolar line in order to validate the current one. The drawback of this method is the complex algorithms involved to recover the pattern. Moreover, the crystal growing procedure does not always lead to the correct identification of the slits, so uncertainty should be considered. The advantage of the method is the possibility of obtaining shape from moving objects. This work is of great value, since it inspired a series of works which dealt with more evolved techniques the problem of lost slits or disorders among slits.

Chen et al. presented a range sensor based on two cameras and an LCD projector [Chen *et al.*, 1997]. The latter projected a unique pattern in order to ease the search for correspondences along pairs of epipolar lines. However, the technique can also be applied when only one camera is used. The pattern consisted of a series of vertical coloured slits separated by black bands. The slit colours were chosen using a trial and error algorithm in order to find a sequence with low autocorrelation in the Hue component of the slits. The decoding method proposed by Chen et al. was the most developed part of the system. The decoding stage was divided into two parts: the *intra-scanline* search and the *inter-scanline* consistency. Since Chen et al. used two cameras, all the points laying on a line in the first image have their correspondences in a line of the second image. Both lines constitute a pair of epipolar lines. The *intra-scanline* process tries to match every edge of every pair of epipolar lines, in order to triangulate their $3D$ position at a later stage. Every epipolar line contains a sequence of colours separated by black gaps corresponding to the projected coloured slits that are visible from the point of view of the camera. Since the cameras in this system have different points of view, for a given pair of epipolar lines, the observed sequences of colours can differ. Therefore, usually one of the cameras perceives more slits than the other one and, therefore, the number of edges differs. In order to match the edges observed by both cameras *dynamic programming* was used. Dynamic programming is capable of obtaining the optimal set of insertions, deletions and substitutions that must be applied to the perceived sequence in order to obtain the original projected sequence. Nevertheless, this algorithm is only robust against deletions and erroneous identification of the colour of some slits, but does not ensure a good solution if disorders among slits have occurred. Therefore, the measuring surface must be monotonic. Furthermore, in the *inter-scanline consistency* stage, an attempt is made to match the edges that have not been matched in the pair of images by using the information of adjacent epipolar lines.

The weak point of the work from Chen et al. is the lack of robustness of the pattern used. Although the sequence of colours is generated accomplishing some constraints, it is not optimal. Furthermore, the authors assumed that the observed slits cannot be reordered with respect to the projected ones, but this is not always true.

2.5.2 Strategies based on De Bruijn sequences

The pattern projection techniques presented in the previous subsection were usually generated by brute-force algorithms in order to obtain some desirable characteristics. In this subsection, a group of patterns encoded with a well-defined type of sequences called *De*

Bruijn sequences is presented. First, a theoretical introduction into the field is given in order to explain why these sequences are suitable for encoding patterns.

A *De Bruijn sequence* of order m over an alphabet of n symbols is a circular string of length n^m that contains each substring of length m exactly once. Similarly, a *pseudorandom sequence* or a *m-sequence* has a length of $n^m - 1$ because it does not contain the substring formed by 0's [MacWilliams and Sloane, 1976]. This sort of sequence can be obtained by searching *Eulerian circuits* or *Hamiltonian circuits* over different kinds of *De Bruijn graphs* [Fredricksen, 1982]. For example, in the graph shown in Figure 2.6a, all the words of length $m - 1$ (with m equal to 4) are included in the vertices. An Eulerian circuit is a path starting and ending in the same vertex and passing through all the edges exactly once. Gathering the edge labels of such circuit, a De Bruijn sequence of order m is obtained

$$1000010111101001 \qquad (2.5)$$

If a Hamiltonian circuit is searched over the same graph (a path which passes through all the vertices only once and starts and ends in the same vertex), a De Bruijn sequence of order $m - 1$ is obtained

$$00101110 \qquad (2.6)$$

An interesting property of a De Bruijn sequence is that it presents a flat autocorrelation function with a unique peak at moment 0. It can be shown that this is the best autocorrelation function that can be achieved and this means that it is clearly uncorrelated. Pseudorandom sequences have been used to encode patterns based on column or row lines and grid patterns. In the following paragraphs, some relevant works using De Bruijn sequences to encode patterns are explained.

Hügli et al. [Hügli and Maître, 1989] improved the pattern proposed by Boyer and Kak [Boyer and Kak, 1987]. In this case, a pattern composed of horizontal coloured slits was also projected. However, the sequence of colours was chosen using a pseudorandom sequence. The authors studied sequences where two consecutive slits with the same colours were not allowed. Therefore, the length of a sequence accomplishing this constraint is $Q(Q - 1)^{N-1}$, where Q is the number of colours used and N the window size.

Monks et al. [Monks and Carter, 1993] designed a pattern based on horizontal coloured stripes in order to reconstruct dynamic scenes. A total number of 6 colours were used to paint the slits, separated by black bands. The colouring of the slits was chosen so that every subsequence of three colours appeared only once. Therefore, a De Bruijn sequence of order three based on an alphabet of six symbols was used. The given sequence was taken from the article published by Hügli and Maitre [Hügli and Maître, 1989].

In this technique, the camera image is thresholded in the HSV colour space, using the Hue component to distinguish among the six colours used, i.e. red, green, blue, yellow, magenta and cyan, which are equally spaced with respect to this component. In the decoding stage, Monks et al. faced the problem of loss of slits for every column of the camera image. Due to surface discontinuities, some of the slits were not observed

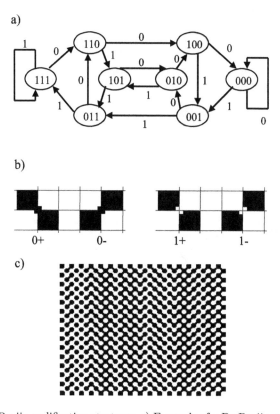

Figure 2.6: De Bruijn codification strategy: a) Example of a De Bruijn graph to construct De Bruijn sequences; b) primitives proposed by Vuylsteke and Ooesterlick to represent the binary values; c) the resulting pattern using the primitives.

by the camera. To recover the position in the pattern of a given slit is necessary to correctly identify the colours of the slit itself and the slits above and below it. The authors decided to build a graph for the whole camera image, where every node represents an image edge between a coloured slit and the black band above it. In every node, the colour of the corresponding slit is stored. Two nodes are connected if the corresponding slits are consecutive in an image column and the distance between them is not very long (otherwise, an occlusion had occurred and some slits may have been deleted). Every column in the image produces a new path in the graph. All the nodes of the graph corresponding to a set of at least three consecutive slits are shared for all those image columns which detected the same subsequence. These nodes are shared since their position in the pattern is directly found thanks to the window property of the De Bruijn sequence. For all the other slits detected in an image column, a new node is inserted in the graph. A minimum-cost, matching algorithm is used to match the original projected sequence and the graph. The match algorithm minimises a cost function based on the costs associated with a node insertion and deletion and the cost of replacing the colour of a node. The system was applied to speech recognition, projecting the pattern to the speakers's face and reconstructing the mouth pose in order to recover the pronounced letters. The work by Monks et al. has a robust decoding stage, which can be applied to other systems based on De Bruijn sequences.

Vuylsteke and Oosterlinck [Vuylsteke and Oosterlinck, 1990] presented a binary encoded pattern by means of De Bruijn sequences. A total number of 63 columns were encoded in a unique pattern, so the system is suitable for moving scenes. The pattern structure is a checkerboard where the column of every grid point is encoded. The encoding system is based on two binary *pseudorandom sequences* of order 6 and length 63, shown in eq. 2.7

$$\{c_k\} = 1111110000010000110001010011110100011001001011011011001101010$$
$$\{b_k\} = \{c_{k-17}\}$$
(2.7)

where $\{b_k\}$ is the same sequence c_k shifted by $\varphi = 17$ positions. Both sequences have a window property of length 6. When combined in a bitmap, as shown in eq. 2.8 for every row, the obtained code assignment has the window property of 2×3. Every element of the sequences represents the individual representation of each grid point of the pattern.

$$
\begin{array}{ccccccccccc}
c_0 & b_1 & c_2 & b_3 & c_4 & b_5 & c_6 & b_7 & c_8 & b_9 & \ldots \\
b_0 & c_1 & b_2 & c_3 & b_4 & c_5 & b_6 & c_7 & b_8 & c_9 & \ldots \\
c_0 & b_1 & c_2 & b_3 & c_4 & b_5 & c_6 & b_7 & c_8 & b_9 & \ldots
\end{array}
$$
(2.8)

In fact, since only column codification is desired, the windows in eq. 2.9 should produce the same codeword.

$$\begin{bmatrix} b_3 & c_4 & b_5 \\ c_3 & b_4 & c_5 \end{bmatrix} \quad \begin{bmatrix} c_3 & b_4 & c_5 \\ b_3 & c_4 & b_5 \end{bmatrix} \tag{2.9}$$

In order to do this, the binary codewords are generated by reading first the elements of c_k and then the elements of b_k. In the example given, the codeword for both windows would be $((c_3, c_4, c_5), (b_3, b_4, b_5))$. To be able to distinguish among elements of c_k and elements of b_k, it is necessary to chose different representations in the pattern. For this purpose, every grid point is marked with a bright or a dark spot representing the binary states 1 and 0. Then, the four neighbouring squares of the grid point in the checkerboard are painted depending on whether the corresponding sequence element belongs to c_k or to b_k. Both representations are shown in Figure 2.6b. The binary states for sequence c_k are labelled 0+ and 1+, while the ones for sequence b_k are 0− and 1−. The whole pattern representation is shown in Figure 2.6c.

The segmentation of the pattern in the camera images is easily done as only two intensity levels are used and also there is symmetry around every grid point. The decoding stage consists in recovering every 2×3 window and obtaining its codeword, which leads to the column position in the pattern. According to the authors, using rectangular windows is more robust than using single row windows, as the neighbourhood involved is more compacted and less sensitive to surface discontinuities. Pajdla reimplemented the whole method, improving the calibration process [Pajdla, 1995].

Later, Salvi et al. proposed a pattern consisting of a grid of thin vertical and horizontal slits [Salvi *et al.*, 1998]. The authors argued that grid coding is a better solution because the grid points can be easily segmented. Furthermore, the neighbourhood around a grid point can be found by just tracking the edges of the grid. The pattern was designed as a 29×29 grid using three different colours for horizontal slits and three more colours for vertical slits. The colour assignment was made by using the same De Bruijn sequence of order 3 (size of window property) for both rows and columns of the grid. The grid intersection points are reconstructed after decoding their position in the pattern using the window property. Some years later, Petriu et al. [Petriu *et al.*, 2000] proposed a similar pattern of 15×15 grid points. However, they did not propose reconstructing the grid crossing points, but reconstructing the four corner points of every intersection. Therefore, the resolution of the system is larger. The only requirement is to increase the thickness of the slits, so that the four corners of a crossing do not fall in the same pixel of the image. Moreover, Lavoie et al. [Lavoie *et al.*, 1999] proposed a similar pattern to the one proposed by Salvi et al. A pseudorandom sequence of 3rd order based on 5 colours was used, obtaining a sequence of length 124 where every subsequence of length 3 is unique. Both vertical and horizontal slits of the grid are coded with the same pseudorandom sequence. The most interesting aspect of the technique proposed by Lavoie et al. is that they do not reconstruct the crossing points, but rather, the curves. For this purpose, *non uniform rational Bézier splines* (NURBS) were used. The NURBS have some nice properties under affine transformations as they are invariant under scaling, rotation, translation, shear and parallel and perspective projection. Once the grid is segmented in the camera image, the recovered grid points are used as control points to interpolate $2D$ NURBS in the image for both rows and columns of the grid. Due to the projection invariance property of a NURBS curve, a reverse projection transform can be performed in order to obtain the 3D NURBS that fit the measuring surface.

Recently, Zhang et al. [Zhang *et al.*, 2002] developed a technique based on De Bruijn codification that achieves excellent performance. The proposed pattern consisted of 125 vertical slits coloured by using a De Bruijn sequence of 3rd order and 5 colours. Zhang et al. studied the problems that occlusions and discontinuities in the measuring surface can produce when observing the projected pattern. As other authors had previously pointed out, they agreed that deletions and disorders among the slits may appear in the observed sequence. In order to match the observed sequence with the projected one, dynamic programming was adopted as in the work by Chen et al. However, Zhang et al. pointed out that simple dynamic programming is only successful when the measuring surface is monotonic, i.e. disorders among slits cannot appear. In order to eliminate such a limitation, they invented the multi-pass dynamic programming. Zhang et al. also extended their technique to the time-multiplexing paradigm by projecting the pattern several times by shifting it consecutively and locating the slits with sub-pixel accuracy in each iteration. Therefore, the total resolution is increased.

2.5.3   Strategies based on M-arrays

There is a set of authors who have adopted the theory of *perfect maps* in order to encode a unique pattern, taking advantage of the interesting mathematical properties of these matrices. In the following paragraphs we give an introduction to this mathematical theory.

Let M be a matrix of dimensions $r \times v$ where each element is taken from an alphabet of k symbols. If M has the window property, i.e. each different submatrix of dimensions $n \times m$ appears exactly once, then M is a *perfect map*. If M contains all submatrices of $n \times m$ except the one filled by 0's, then M is called an *M-array* or *Pseudorandom array*. For more information see [MacWilliams and Sloane, 1976] and [Etzion, 1988]. This kind of array has been widely used in pattern codification because the window property allows every different submatrix to be associated with an absolute position in the array.

M-arrays can be constructed by folding a pseudorandom sequence. In order to create an M-array of dimensions $n_1 \times n_2$, a pseudorandom sequence of length $n = 2^{k_1 k_2} - 1$ is required, where $n = n_1 \cdot n_2$, $n_1 = 2^{k_1} - 1$ and $n_2 = n/n_1$. The resulting array has a window property of $k_1 \times k_2$. The procedure is as follows: the first element of the sequence is placed in the north-west vertex of the array. Successive elements of the sequence are written in the array following the main diagonal and continuing from the opposite side whenever an edge is reached. For example, given the binary pseudorandom sequence of the 4th order 000100110101111 whose length is $n = 15$, good parameters for constructing an *M-array* are $n_1 = 2^2 - 1 = 3$ and $n_2 = 15/3 = 5$. Then, an *M-array* of 3×5 with window property 2×2 is obtained. It must be noted that *M-arrays* also share the *circular* property of the pseudorandom sequences, so it is necessary to complete the array by adding to the right the firsts $k_2 - 1$ columns and the firsts $k_1 - 1$ rows below it. The complete array is shown in eq. 2.10

$$
\begin{bmatrix}
0 & 1 & 1 & 1 & 1 & 0 \\
0 & 0 & 1 & 1 & 0 & 0 \\
0 & 1 & 0 & 0 & 1 & 0 \\
0 & 1 & 1 & 1 & 1 & 0
\end{bmatrix}
\tag{2.10}
$$

The main differences between the techniques included in this group is the way in which the elements of the array are represented in the pattern. Some authors prefer to define the pattern as an array of coloured spots, where each colour represents one of the symbols of the coding alphabet. Other authors prefer to define different shapes for each symbol. When perceiving the projected pattern, an algorithm to recover the maximum number of visible windows must be fulfilled. This is the crucial step of these systems. Since spatial neighbourhood is used, not all the pattern will be visible from the camera's point of view, due to shadows and occlusions. The robustness of these methods depends on the correct decodification of the visible parts, taking advantage of the properties of the M-arrays.

Using arrays to codify a pattern means that a bidimensional coding scheme is being used, because every point of the pattern has a different codeword which encodes both vertical and horizontal coordinates. Since the codification is concentrated in one pattern, these techniques are suitable for measuring dynamic scenes. However, some authors prefer to project additional patterns in order to ease the segmentation part of the system or to carry out an intensity or colour normalisation. In this case, the system is limited to static scenes. In any case, the number of projected patterns is always lower when compared to time-multiplexing methods. In the following paragraphs, the existing patterns based on M-arrays are addressed and the most relevant are briefed.

A binary M-array of 24×24 was proposed in 1988 by Morita et al. [Morita *et al.*, 1988]. This array has the window property of 3×4. The M-array representation is made by painting black dots on a white background, for the array elements corresponding to symbol 1. Two patterns are projected on the measuring surface: the first one contains all the possible black dots in order to locate their centres in the camera image. The second pattern is the M-array representation. Therefore, the method is restricted to static scenes. However, it can be adapted to moving scenes by only projecting the M-array pattern and making the segmentation and decoding algorithm more robust.

Petriu et al. [Petriu *et al.*, 1992] used an M-array to encode a grid pattern where each cross-point represents an element of the M-array. The binary state of every cross-point is represented with the presence or absence of a square painted on it. The system was intended for object recognition, based on a database containing previously reconstructed surfaces.

Some years later, Kiyasu presented an interesting study [Kiyasu *et al.*, 1995]. The aim of the work was to obtain the shape of specular polyhedrons, i.e. objects composed of flat surfaces with high reflectance. Normally, most coded structured light systems are not intended to reconstruct specular surfaces but lambertian ones. A binary M-array represented with a grid of 18×18 circular spots with a window property of 4×2 was used.

Spoelder et al. began to develop a prototype to measure the shape of the cornea [Spoelder *et al.*, 2000], by means of projecting a binary M-array of 65×63 elements. Cyan and yellow were used to encode the binary values of the M-array. The structure of the designed pattern is a checkerboard, where the white squares are used to place the elements of the M-array. In Figure 2.7a a portion of the pattern can be observed. The black squares were introduced in order to ease the pattern segmentation in the camera images. Due to the complex reflectance characteristics of the cornea, the recovered pattern from the camera images has a lot of data loss. This required the design of a complex segmentation algorithm, which we will now summarise. Firstly, the cross-points of the checkerboard are located by mask filtering. Secondly, every detected cross-point is labelled by observing the colours of the adjacent non-black squares and using the window property. Then, a graph is constructed by linking the neighbours. This step leads to a series of disconnected *subpatterns* that must be matched to the original projected pattern. Each subpattern is positioned on the projected pattern in the position where the minimum Hamming distance is achieved. The elements which do not fit in the original pattern are intended to be corrected.

One of the most famous works of this group is due to Griffin et al. [Griffin *et al.*, 1992]. The authors defined a systematic process for constructing a maximum size array of $n \times m$ based on an alphabet of b symbols with certain restrictions: every element of the array has a unique codeword formed by its own value and the values corresponding to its four neighbours (north, south, east and west). As can be seen, such an array is a special case of *perfect maps*, since it has window property of 3×3, but not all the possible windows appear. Some authors call these arrays *perfect submaps*. The construction process of such matrices is as follows: first, let Vhm be the sequence based on alphabet b containing all the possible triplets of symbols (a *De Bruijn sequence*). Let Vvm be the vector made by the sequence of all the pairs of symbols of alphabet b. Consequently, the first row of the matrix is $f_{0i} = Vhm_i$.

The rest of the matrix elements are calculated using eq. 2.11. The row index is indicated with i and varies from 0 to the length of Vhm, while j is the column index varying from 0 to Vvm length.

$$f_{ij} = 1 + ((f_{i-1j} + Vvm_j) \; mod \; b) \tag{2.11}$$

For example, if an alphabet $b = \{1,2,3\}$ is taken, then the following vectors are obtained

$$Vhm = (33132131123122121113323222333)$$
$$Vvm = (3121132233) \tag{2.12}$$

Then, applying the algorithm the matrix shown in Figure 2.7b is generated.

For their experiments, Griffin and Yee generated an array of 18×66 using the alphabet of four symbols $\{1, 2, 3, 4\}$. In order to project this array, two strategies were adopted. The first consisted of representing each alphabet element with a different colour. Then the projected pattern was defined as an array of coloured dots. The second approach consisted of defining a set of shape primitives for every element of the alphabet. An example of such primitives is shown in Figure 2.7b. Then the background of the pattern is painted black

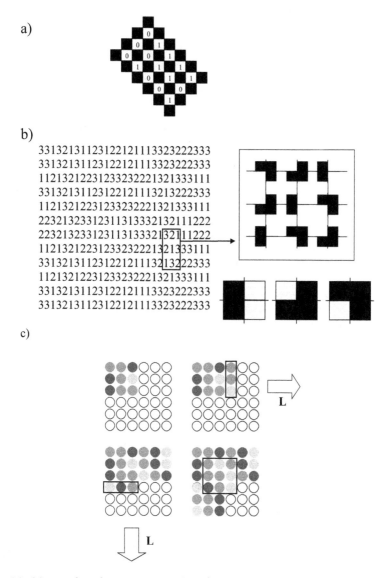

Figure 2.7: M-array based patterns examples: a) Binary M-array located in a checkerboard (the 0 and 1 symbols are replaced by two different filling colours); b) Example of M-array based on 3 symbols proposed by Griffin et al. Three shape primitives were proposed to represent the symbols of the alphabet {1,2,3}; c) Morano et al. algorithm to generate M-arrays with coloured spots representation.

and white forming a grid of 18×66 with one of the primitives at every crossing point. The representation of a window of 3×3 of the M-array is shown in Figure 2.7b. This second representation of the M-array is much more robust in the presence of coloured objects. Some years later, Hsieh presented an analytical method for decoding the position of a given codeword of Griffin's array [Hsieh, 2001], using just simple arithmetic operations with the elements of the window and decoding the pattern quickly.

Davies and Nixon [Davies and Nixon, 1998] proposed a unique pattern of coloured spots for obtaining shapes from dynamic scenes. Specifically, the system was applied for automatic speech identification by projecting the pattern onto the speakers' face at video rate. The spots are coded by following Griffin's method. Cyan, yellow and magenta colours were chosen to paint the spots, which are placed hexagonally in the pattern. In this technique a segmentation algorithm is applied for obtaining the image coordinates of the visible dots. First, an edge detector filter is used in order to find the contours of the perceived ellipses corresponding to the projected circular dots. For every epipolar line in the camera image, all the ellipses nearly positioned onto the line are searched. Then accurate positions of the ellipses are found using an adapted formulation of the Hough Transform. When all possible ellipses have been located, the decoding process using the window property leads to correspondences between the camera image and the projected pattern.

One of the most interesting techniques from this group was given by Morano et al. [Morano et al., 1998]. The authors proposed an algorithm for constructing an M-array, fixing the length of the alphabet, the window property size, the dimensions of the array and the Hamming distance between every window. Previously all the methods worked with a Hamming distance of one, which did not allow error correction. In fact, the arrays used by Morano et al. are simply *perfect submaps* since not all the possible windows are included.

The algorithm used to generate an array with fixed properties is based on a brute-force approach. For example, when constructing an M-array based on three colours with window property of 3×3 the following steps are taken: first, a subarray of 3×3 is chosen randomly and is placed in the north-west vertex of the M-array that is being built. Then consecutive random columns of 1×3 are added to the right of this initial subarray, maintaining the integrity of the window property of the array and the Hamming distance between windows. Afterwards, rows of 3×1 are added beneath the initial subarray in a similar way. Then, both horizontal and vertical processes are repeated by incrementing the starting coordinates by one, until the whole array is filled. Whenever the process reaches a state where no possible elements can be placed, while accomplishing the global window property, the array is cleared and the algorithm starts again with another initial subarray. The basic steps of the algorithm are represented in Figure 2.7c. The study of the performance of this algorithm showed that using M-arrays of 45×45 pixels with window property of 3×3 using 3 or more colours, fixed Hamming distances between windows from the typical 1 up to 4 can be generated. Moreover, in most cases, multiple solutions can be found.

Once the generated pattern is projected onto the measuring surface, it must be recovered and the dots must be well labelled in order to find correspondences between the camera and the projector. As every dot is contained in 9 windows, the authors applied a voting algorithm where every window proposes a codeword of length 9 (which indicates its position in the pattern) for every one of its elements. Then every observed dot has up to 9 codewords proposed by every window to which it belongs. The codeword with the maximum number of votes is the more reliable, so it is used to label the dot. The results showed that when using a Hamming distance of 3 instead of 1, the number of dot mislabelings decreases, due to the possibility of correcting one error per window.

Another interesting contribution made by Morano et al. was to note that a system based on M-arrays can also be used when colour cannot be projected (because the scene is too colourful or because a colour camera is not available). If N colours are used to encode the M-array, $log_2(N) + 1$ patterns can be projected, encoding every colour with a binary codeword. The system becomes more robust since only two intensity levels are used but it is limited to static scenes.

2.6 Direct codification

There are certain ways of creating a pattern so that every pixel can be labelled by the information represented on it. Therefore, the entire codeword for a given point is contained in a unique pixel. In order to achieve this, it is necessary to use either a large range of colour values or introduce periodicity.In theory, a high resolution of $3D$ information can be obtained. However, the sensitivity to noise is very high because the "distance" between "codewords", i.e. the colours used, is nearly zero. Moreover, the perceived colours depend not only on the projected colours, but also on the intrinsic colour of the measuring surface. This means, in most cases, that one or more reference images must be taken. Therefore, these techniques are not typically suitable for dynamic scenes. Direct codification is usually constrained to neutral colour or pale objects. For this reason, it is necessary to perceive and identify the whole spectrum of colours, which requires a "tuning" stage that is not always easy to achieve (depending on the devices used).

We shall now discuss two groups of methods using direct codification: a) codification based on grey levels: a spectrum of grey levels is used to encode the points of the pattern; b) codification based on colour: these techniques take advantage of a large spectrum of colours.

2.6.1 Codification based on grey levels

Carrihill and Hummel [Carrihill and Hummel, 1985] developed a system called *intensity ratio depth sensor*. It consists of a linear wedge spread along vertical columns containing a scale of grey levels. A ratio is calculated between every pixel of the perceived wedge and the same pixel value under constant illumination. This ratio is related to the column of the pattern that has been projected in the pixel. Since two patterns must be projected dynamic scenes are not considered. The authors used a slide projector and a monochrome camera with 8 bits of intensity per pixel. The authors aimed to tune the setup so that the relationship between the ratio and the image column number was nearly linear. However, Carrihill and Hummel achieved poor accuracy in their measurements, with a mean error of about 1 *cm*. This was due to the high sensitivity to noise and non-linearities of the projector device.

Figure 2.8: Pattern proposed by Carrihill and Hummel

Miyasaka et al. [Miyasaka *et al.*, 2000] reproduced the intensity ratio depth sensor by using an LCD projector and a 3CCD camera. With this setup, more accurate results were obtained. The authors took into account that the reflectance of the surface points is not constant for all the light frequencies and each RGB channel of the camera was treated independently. Furthermore, only a narrower band of light frequencies was considered.

Chazan and Kiryati [Chazan and Kiryati, 1995] carried out experiments using an extension of the Carrihil and Hummel method called *pyramidal intensity-ratio depth sensor*, also known as the *sawtooth sensor*. The motivation behind this new approach was the high sensitivity to noise of the original method. As a wide intensity spectrum is projected in only one shot, the camera must be able to perceive such a spectrum nearly linearly, which is very difficult to achieve using an LCD projector. The new method consisted of consecutively projecting the linear wedge by increasing its period. Therefore, the first pattern is a simple-period wedge from black to white. The second contains two linear wedges, the third contains four wedges and so on. At the end, the last pattern contains 2^n linear wedges. Since every period is a linear wedge from black to white, the last pattern uses less grey levels in each period. This means that adjacent grey levels in the last pattern are less similar and easily distinguishable. However, since periodicity is present, the grey level of a certain pixel in the last perceived pattern is not enough to decode its position. To resolve the ambiguity the previously viewed patterns are used. This strategy is quite similar to the time-multiplexing techniques, but in this case, the exact codewords are not recovered. Moreover, since the sharp transitions between periods can lead to high errors, every periodic pattern is projected twice, shifting it by half a period. Then when reading grey levels close to a period transition (black or white), the corresponding shifted pattern is used to avoid the period transition. For every projected pattern an image of the scene is grabbed. Then an intensity ratio is calculated for every image with respect to a constant light image. The *sawtooth sensor* is more accurate than the classic *Intensity Ratio Depth Sensor*. Experiments made by the authors over distances of about 80 *cm* show that the typical errors of 1 *cm* of the *Intensity Ratio Depth Sensor* can be reduced to 1 *mm* with the *Pyramidal Intensity Ratio Depth Sensor*. However, the number of patterns increases substantially.

Prior to Chazan and Kiryati's work, Hung proposed a grey level sinusoidal pattern [Hung, 1993]. The author pointed out that the period of the observed pattern increases proportionally with the distance between the projector and the object, and therefore, the frequency decreases. The idea was to estimate the instant frequency in every pixel of the camera image and then depth can be calculated for every pixel.

The system was tested with synthetic images with gaussian noise. Although the results were good, real experiments should be completed taking into consideration the non-linear behaviour of the devices. Furthermore, since the pattern is periodic, ambiguity problems can arise.

2.6.2   Codification based on colour

The methods belonging to this group use the same principle as the ones discussed in subsection 2.6.1. However, colour is used to encode pixels instead of using grey levels. For instance, Tajima and Iwakawa [Tajima and Iwakawa, 1990] presented the rainbow pattern. A large set of vertical narrow slits were encoded with different wavelengths, so that a large sampling of the spectrum from red to blue was projected. In order to project this spectrum, a nematic liquid crystal was used to diffract white light. The images were grabbed by a monochromatic camera with 11 bits of intensity depth. Two images of the scene were taken through two different colour filters. By calculating the ratio between both images an index for every pixel is obtained that does not depend on illumination, nor on the scene colour. Geng [Geng, 1996] improved on this approach by using a CCD camera and a linear variable wavelength filter in front of it. Hence, only a single image of the measuring surface had to be captured from the scene.

Sato presented the *multispectral pattern projection range finder* [Sato, 1999]. In this work, the author discussed the complicated optical system required for the rainbow range finder of Tajima. Sato proposed a new technique that only needs an LCD projector and a CCD camera. Moreover, the new technique could eliminate the colour of the measuring surface, so the results were not affected by the spectral reflectance of the surface. The technique consisted of projecting a periodic rainbow pattern 3 times, shifting the hue phase 1/3 of its period in every projection. An extra image was synthesised by a certain linear combination of the three grabbed images. Afterwards, Sato demonstrated that the Hue value of every pixel of the synthesised image is equal to the projected Hue value in the first pattern. Therefore, correspondences between synthesised image pixels and projected rainbow columns can be done. In order to get a good resolution, the pattern had to be periodic, so the identification of the periods is a key point in the decoding stage.

Wust and Capson presented a technique based on a three-step phase shifting [Wust and Capson, 1991]. However, instead of projecting three times a periodic pattern shifted in every projection, a single pattern was used. The pattern was designed with three overlapping sinusoids shifted between them, in order to encode the columns of every row. The first sinusoid is represented with red, the second is shifted $90°$ and is represented with green, and the third, which is shifted $90°$ with respect to the green one, is represented with blue. Then, once the pattern is projected and an image is grabbed, the phase shift can be calculated for every pixel using the following equation

$$\Phi(x, y) = \arctan\left(\frac{I_r - I_g}{I_g - I_b}\right) \tag{2.13}$$

$\Phi(x, y)$ is the phase in a given pixel where the intensities of the red, green and blue are denoted as I_r, I_g and I_b respectively. The technique of Wust and Capson requires only a unique pattern, so moving surfaces can be measured. However, the surface must be predominantly colour neutral and must not contain large discontinuities.

2.7 Experimental results

We implemented a set of 7 representative techniques taken from the proposed classification groups. All the techniques have been tested under the same conditions in order to evaluate their advantages and constraints. A shape acquisition framework as the typical application of coded structured light has been used. This enables us to test the different techniques in terms of number of correspondences and how accurately they are found.

A low-cost structured light system was used. It is composed of an LCD video-projector (Mitsubishi XL1U) working at 1024 × 768 pixels, a camera (Sony 3CCD) and a frame grabber (Matrox Meteor-II) digitising images at 768 × 576 pixels and 3 × 8 bits per pixel. A standard PC was used for implementing the algorithms. For the mathematical details concerning the calibration of cameras we refer to the comparative review presented by Salvi et al. [Salvi et al., 2002]. This review concludes that non-linear models which take into account the radial distortion introduced by the lenses are enough to obtain high accuracy. In our case, an adaptation of the Faugeras calibration method which includes radial distortion has been used [Faugeras, 1993].

Some calibration methods use a set of coplanar 3D points [Batista et al., 1999] while others use a set of non-coplanar points, which is our case. Our calibration procedure requires a set of non-coplanar 3D points and its corresponding 2D projections. To obtain the sample points, we used two orthogonal white panels each one containing 20 black squares, placed about 1 meter in front of the camera. The world frame is positioned at the intersection of both panels as shown in Figure 2.9. Then, the camera captures an image and the 2D corners are detected with sub-pixel accuracy by means of image processing. Finally, 2D and 3D points are used to iterate the calibration algorithm to convergence.

The LCD projector can be modelled as an inverse camera, so that the same camera model remains valid. In order to calibrate the projector, we assume that the camera has been previously calibrated and that the calibrating panels remain in the same position. The projector is placed aside the camera so that the angle between the optical axis of both devices is about 15 degrees. A white grid pattern is projected onto both panels. We chose the grid cross-points of the pattern as 2D points. Then, it is necessary to capture an image with the camera. The grid cross-points can be detected with sub-pixel accuracy. Afterwards, since the camera is already calibrated, the 3D points can be reconstructed,

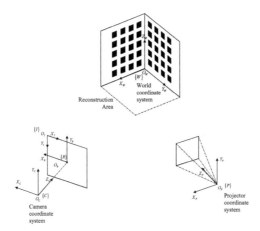

Figure 2.9: Geometric calibration setup.

taking into account certain geometrical constraints. Remember that the grid is projected onto the calibration panels and the world frame is positioned on the bottom angle of both panels, as shown in Figure 2.9. Then, the left panel lies on the plane $Y = 0$, while the equation of the plane containing the right panel is $X = 0$. In this way, the $3D$ points, corresponding to the grid cross-points projected onto the panels, can be triangulated by intersecting the camera rays with the equation of these planes. Then, the set of $2D$ and $3D$ points obtained with this procedure are used to calibrate the projector's model.

The implemented techniques are listed below and are represented in Figure 2.10.

- Time-multiplexing:
    - Posdamer: stripe patterns encoded with Gray code of 7 bits so that 128 stripes are encoded [Posdamer and Altschuler, 1982].
    - Gühring: the *line shifting* technique using 6 Gray coded patterns and 21 slits shifted 6 times [Gühring, 2001].
    - Horn: three patterns encoding 64 stripes by using 4 grey levels [Horn and Kiryati, 1999].

- Spatial neighbourhood:
    - De Bruijn: a pattern with 64 vertical slits encoded with a *De Bruijn* sequence of 3rd order and 4 colours.
    - Salvi: a grid pattern of 29×29 slits encoded with a De Bruijn sequence of 3rd order and 3 colours [Salvi *et al.*, 1998].
    - Morano: pattern consisting of colour dots encoded with an M-array of 45×45 elements and 3 colours [Morano *et al.*, 1998].

- Direct codification:

    - Sato: the three periodic patterns proposed by Sato [Sato, 1999] were employed.

Some images showing the compared patterns illuminating one of the test objects are presented in Figure 2.11.

The performance of the techniques was evaluated by means of quantitative and qualitative tests. In the following subsections the experiments and their results are presented.

2.7.1   Quantitative evaluation

A white plain (flat surface) at a distance of about $120cm$ to the camera was reconstructed 30 times using all the implemented techniques. A multiple regression was applied in order to obtain the equation of the $3D$ plane for every technique and for every reconstruction. The same experiment was repeated by bringing the plain closer to the camera by about $40mm$. Then the average and the standard deviation of the distance between both plains was calculated for every technique. The results of the experiment are shown in table 2. The table includes the standard deviation, in μm, of the average distance between both parallel plains, the average number of $3D$ points that were reconstructed, the % of image pixels inside a region of 515×226 pixels that were decoded, and the total number of projected patterns for every technique (including white and black patterns for intensity normalisation when needed).

Table 2.2: Quantitative results. The headings are: author's name of the technique; standard deviation of the reconstructing error; average number of $3D$ points; % of pixels from images that have been reconstructed in average; number of projected patterns.

Technique	Stdev (μm)	3D Points	Resolution %	patterns
Posdamer	37.6	25213	21.67	9
Horn	9.6	12988	11.17	5
Gühring	4.9	27214	23.38	14
De Bruijn	13.1	13899	11.94	1
Salvi	72.3	372	0.32	1
Morano	23.6	926	0.80	1
Sato	11.9	10204	8.77	3

2.7.2   Qualitative evaluation

In order to evaluate the performance of the techniques, it is also useful to observe the reconstruction of certain surfaces and analyse them from a qualitative point of view. For this purpose, two surfaces were reconstructed.

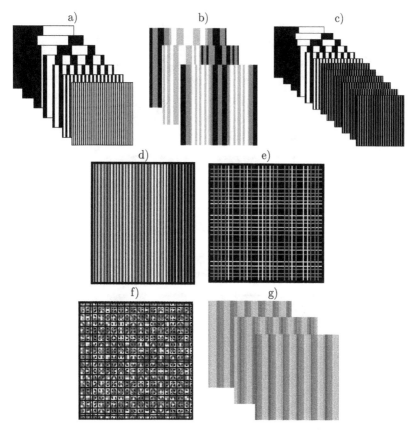

Figure 2.10: Patterns corresponding to the implemented techniques: a) Posdamer; b) Horn and Kiryati; c) Gühring; d) De Bruijn; e) Salvi; f) Morano; g) Sato.

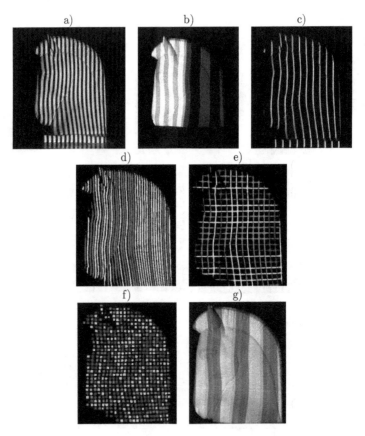

Figure 2.11: Some of the patterns illuminating a horse statue (the techniques consisting of a sequence only one of the patterns is shown): a) Posdamer; b) Horn and Kiryati; c) Gühring; d) De Bruijn; e) Salvi; f) Morano; g) Sato.

The first surface was a statue of a white horse of dimensions $21 \times 15 \times 10 \ cm$. The statue and the reconstructions obtained are shown in Figure 2.12. The reconstructions are presented both as clouds of points and rendered surfaces. Techniques with higher resolution (time-multiplexing techniques and De Bruijn patterns based on a single axis codification) enable details of the horse's profile to be distinguished, while other techniques with lower resolution (mainly based on spatial neighbourhood) obtain basically the global profile.

The second test consisted of reconstructing a human hand. This surface is useful for evaluating the performance of the techniques when the surface violates monotonicity, i.e. it contains discontinuities. In this case, the discontinuities are produced by the gaps between the fingers. Results are shown in Figure 2.12. Techniques based on time-multiplexing are not affected since for recovering the codewords of a pixel, it is only necessary to gather its value along the projected patterns. Techniques based on spatial neighbourhood using a single axis codification (De Bruijn) suffer large amounts of data loss as the local smoothness assumption of the measuring surface is violated. Nevertheless, techniques that encode both pattern axis (Morano and Salvi) can identify some regions near discontinuities due to the propagation of codewords among adjacent points. Direct coding techniques should be robust against discontinuities if no periodicity is used in the patterns. Since the technique proposed by Sato exploits periodicity, it fails when reconstructing the fingers. Furthermore, periodicity is required for such techniques in order to reduce the number of colours in the pattern as it is very difficult to correctly differentiate among the emitted colours if a large spectrum is used.

The experiments that have been carried out allow comparison of the different groups of techniques classified. It has been shown that techniques based on time-multiplexing achieve the most accurate results. Moreover, line-shifting combined with Gray Code permits exploitation of the whole theoretical resolution of patterns. The results also demonstrate that locating the pattern stripes with sub-pixel accuracy (in the case of Gühring [Gühring, 2001] and Horn [Horn and Kiryati, 1999] implementations), leads to better results than using pixel accuracy (in the case of Posdamer [Posdamer and Altschuler, 1982] current implementation). Techniques based on spatial neighbourhood have also obtained satisfactory results. For example, the pattern consisting of vertical slits coded with a De Bruijn sequence has obtained very accurate measurements as the slits are also detected with sub-pixel accuracy. However, it has failed when measuring discontinuities. Such problems could be partially solved by using dynamic programming [Zhang et al., 2002]. Furthermore, techniques based on both axis codification, i.e. the grid by Salvi et al. [Salvi et al., 1998] and the array of dots by Morano et al. [Morano et al., 1998], are more robust against discontinuities as redundancy in the coding strategy permits extension of decoded regions to contiguous non-decoded regions. Finally, the direct coded pattern presented by Sato [Sato, 1999] has obtained very accurate results (also locating the stripes with sub-pixel accuracy) and robustness against colourful surfaces. However, this technique has the problem of stripe decodification among the pattern due to its periodic structure when a surface containing discontinuities is measured. Such problems could be overcome by projecting some Gray patterns to remove the ambiguity between periods.

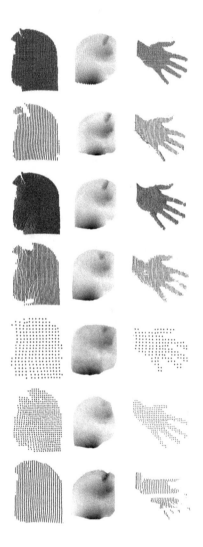

Figure 2.12: Reconstruction results for every one of the implemented techniques. From up to down: Posdamer, Gühring, Horn, De Bruijn, Salvi, Morano and Sato. At left, the cloud of points corresponding to the horse statue reconstruction. In the middle, the corresponding rendered surface from another view point. At right, the cloud of points from reconstruction of a human hand.

2.8 Conclusions

We have presented a comprehensive survey of coded structured light techniques. A new classification of the reviewed techniques has been proposed from the point of view of the coding strategies used to generate the projected patterns.

Time-multiplexing was the first paradigm of coded structured light used to obtain 3D data from an unknown surface. The advantages of these techniques are the easy implementation, the high spatial resolution and the accurate 3D measurements that can be achieved. The main drawback is their inapplicability to moving surfaces since multiple patterns must be projected. Techniques based on projecting stripe patterns encoded with Gray code can obtain very good accuracy, but maximum resolution cannot be achieved. In order to obtain maximum resolution, a technique based on a combination of Gray code and Phase shifting must be used. In this subgroup, the technique proposed by Gühring [Gühring, 2001] must be highlighted. Its drawback however, is the large number of projecting patterns (32 patterns when using maximum resolution). If maximum spatial resolution is not the principal aim of the application, but rather the minimisation of the number of projecting patterns, a technique based on n-ary codes is appropriate. Such methods obtain an accuracy equal to or even better than a Gray code approach, reducing exponentially the number of projecting patterns. For example, a Gray code technique based on the projection of 8 patterns can encode 256 stripes, while an n-ary technique only requires 3 patterns and 13 grey levels or colours to obtain such resolution, for $n = 13$. However, the system using n-ary codes must be calibrated in order to correctly differentiate among the set of grey levels or colours used. If a good calibration cannot be achieved, then it is recommended to reduce the number of grey levels or colours by projecting more patterns.

Spatial neighbourhood coding is the second big group of coded structured light techniques. The advantage compared with time-multiplexing is that such strategy permits, in most cases, moving surfaces to be measured. However, since the codification must be condensed in a unique pattern, the spatial resolution is lower. Moreover, local smoothness of the measuring surface is assumed in order to correctly decode the pixel neighbourhoods. Since this local smoothness is not always accomplished, errors in the decoding stage can arise producing false correspondences. In order to minimise such errors, the algorithms of the decoding stage must be more robust, resulting in an increase in the overall complexity of the technique. Techniques which define the neighbourhoods empirically usually present pattern periodicity or repetition of neighbourhoods, which is not recommended. Such problems have been eliminated by strategies based on De Bruijn sequences and M-arrays.

Techniques based on a unique pattern coded using a De Bruijn sequence have a trade off between the length of the sequence, i.e. the resolution of the system, the number of colours involved and the size of the window property. Most of these methods use either horizontal or vertical windows with a limited size in order to preserve the local smoothness assumption of the measuring surface. If the window size is not too big (in our opinion a good limit is about 10% of the sequence length), more than two colours must be used with the aim of preserving a good resolution. The number of colours used increases the noise sensitivity when measuring colourful scenes. Using up to 6 colours is not very problematic.

With 5 colours and a window size of 3, a resolution of 125 slits per pattern (similar to a Gray code system based on 7 patterns) can easily be achieved with a robust decoding stage. The most complete technique that can be found in the bibliography is the one proposed by Zhang et al. [Zhang *et al.*, 2002]. This technique takes into account that disorders between elements of the sequence can occur when projecting the pattern. The solution proposed is based on multi-pass dynamic programming, which seems the most robust way to recover the original sequence. Furthermore, techniques based on *M-arrays* are more difficult to generate. However, since every coded point has both row and column codewords, a higher degree of redundancy is included. In order to take advantage of this redundancy, an additional step must be programmed in the decoding stage for validating the codeword of every coded point. Similar trade offs to the ones involved when using De Bruijn sequences also appear with M-arrays-based patterns. The segmentation complexity of the observed patterns in such techniques must also be addressed. The most typical representations of an M-array in a pattern are the grid representation and the array of dots. In our opinion, the grid representation can be segmented more easily by edge detection. The encoded points are the intersection of edges, so they can be found very accurately. In addition, when projecting dots, their mass centres must be located. So that it is important to detect when a dot appears partially occluded, since its mass centre will be incorrect. Moreover, the grid techniques allow adjacent cross-points to be located by tracking the edges, while with the dot representation, some sort of euclidian distance must be used to locate the neighbours of a given dot.

It should be noted that a technique based on spatial neighbourhood can always be translated to a time-multiplexed technique by expressing the colours in binary intensity levels distributed over a sequence of patterns.

Direct coding techniques are useful for achieving large spatial resolution and few projecting patterns. However, these techniques present a lot of drawbacks. Firstly, the limited bandwidth of LCD projectors provokes integration of intensities over adjacent pixels. Secondly, variations of light intensities due to the different colours and depths of the measuring surface. Finally, the error quantisation introduced by the camera, which is very sensitive to noise. Therefore, the correct identification of every projected intensity or colour is not easy to achieve. In most cases, the use of such techniques requires a device that projects a unique wavelength for every grey level or colour. Therefore, LCD projectors are not suitable for such a purpose. Some authors use non-standard optical devices to decompose white light, producing monochromatic light planes. Furthermore, since a large spectrum of wavelengths are used, cameras with large depth-per-pixel must be considered (about 11 bits per pixel) for accurate quantisation. It should be noted that most of these techniques cannot measure moving scenes because they need additional patterns to normalise intensity or colour. In addition, these techniques are usually limited to colour-neutral surfaces. Nevertheless, some techniques that can be implemented with an LCD projector and a standard CCD camera were proposed by Wust and Capson [Wust and Capson, 1991] and Sato [Sato, 1999]. Moreover, both techniques are theoretically capable of reconstructing colourful surfaces, and the technique by Wust and Capson can also measure moving surfaces. Accordingly, experimental results given by the technique by Sato showed that, in the synthesised image, most part of the surface colours are eliminated.

Finally, the intensity-depth per pixel used by a coded structured light technique is also an important parameter. The noisier the application environment in which the technique will be applied, the smaller the number of grey levels or colours used should be. Therefore, time-multiplexing techniques based on binary patterns are the most robust against noise. However, when increasing the number of grey levels or colours, differentiating the slits becomes more difficult. The non-linearities of the light spectrum of the projector and the spectral response of the cameras, and the non-uniform albedo of the measuring surface mean that the read colours hardly match with the projected ones. In order to overcome this problem, a full colourimetric calibration procedure should be considered. The illumination model proposed by Caspi et al. [Caspi *et al.*, 1998] or even a simple linear normalisation may be a good solution.

Chapter 3

A proposal of a new one-shot pattern

In this chapter we present a new coloured pattern which is able to obtain correspondences with a unique shot. The pattern is generated by a new coding strategy based on De Bruijn sequences. The new pattern is compared to similar existing patterns by reconstructing the shape of different objects and analysing the results from a quantitative and a qualitative point of view. The results discussed at the end of the chapter show that the pattern is able to increase the number of correspondences in a single shot without loss of accuracy.

3.1 Introduction

As it has been shown in the previous chapter, important efforts have been done in order to generate patterns able to obtain correspondences with a unique projection. As already mentioned, there is an important group of one-shot techniques which are based on coloured multi-slit and stripe patterns. Multi-slit patterns introduce black gaps between the coloured bands of pixels so that two consecutive slits can have the same colour. The black gaps allow intensity peaks to be detected in the images. Every intensity peak corresponds to the central position of a certain slit. Thus, the correspondence problem is solved for points falling onto the centre of every slit which are imaged as intensity peaks. Hereafter, we refer to this matching strategy as *peak-based matching*.

On the other hand, in stripe patterns no black gaps are introduced between the coloured bands of pixels. Therefore, adjacent stripes cannot share the same colour. In such patterns, edges between stripes are searched in the image in order to find correspondences so that an *edge-based matching* is performed.

Note that in multi-slit patterns an edge-based matching is not suitable since a certain amount of the intensity of a slit is integrated over the surrounding black regions. Therefore, the edges in the image do not really correspond to the edges in the pattern. Such problem already appears when multiplexing binary codes by projecting a sequence of binary stripe patterns of successively increasing frequency [Posdamer and Altschuler, 1982]. As stated by Trobina [Trobina, 1995], the best solution in these cases consists in projecting every one of the binary stripe pattern and its negative version. Then, by intersecting the intensity profiles of every pattern pair the edges can be detected with high accuracy. Unfortunately, this solution is not applicable to one-shot techniques. Note also that in the case of coloured stripe patterns a peak-based matching is neither feasible since the pattern is projected with a flat intensity profile.

This chapter proposes a new coloured pattern which combines the advantages of multi-slit and stripe patterns. In Section 3.2 the new coding strategy used to generate the pattern is exposed. Afterwards, details concerning how to segment the pattern and how to solve the correspondence problem are explained in Section 3.3. Then, Section 3.4 presents two calibration procedures for improving the accuracy and robustness of a structured light setup using coloured patterns like the one proposed. Experimental results of the new pattern compared to other similar patterns are shown and discussed in Section 3.5. The chapter ends with conclusions concerning the new approach.

3.2   The new coding strategy

This section presents a new type of coding strategy which defines patterns containing both edges and intensity peaks. The new hybrid pattern appears as a stripe-pattern in the RGB space while in the intensity channel it appears as a peak-based pattern. The intensity profile of the pattern is defined as a square function so that it alternates from a medium intensity stripe to a high intensity stripe.

The coding strategy requires only $n = 4$ different hue values in order to colour a pattern with 128 stripes so that a window property of $m = 3$ stripes is obtained. The formal definition of the pattern can be noted as follows. Let W_m and W_h be respectively the width in pixels of a medium-intensity stripe and a high-intensity stripe. Let $S = \{1, .., 2n^m\}$ be the set of stripe indices. Then, the mapping from the pattern abscissas to the stripe indices is defined by the following function

$$stripe : X \longrightarrow S$$
$$stripe(x) = 2\left((x{-}1) \div (W_m{+}W_h)\right) + \begin{cases} 1 & ((x{-}1) \bmod (W_m{+}W_h)){+}1 \leq W_m \\ 2 & \text{otherwise} \end{cases} \qquad (3.1)$$

where $\div$ denotes integer division. The following function determines the intensity level of a given stripe, which has two possible values $L = \{high, medium\}$

$$Int : S \longrightarrow L$$
$$Int(x) = \begin{cases} high & x \bmod 2 = 0 \\ medium & \text{otherwise} \end{cases} \qquad (3.2)$$

The pattern is divided in n periods $P = \{1, .., n\}$ of n^m consecutive stripes each one.

In order to determine to which period belongs to a given stripe the following function is defined

$$period : S \longrightarrow P$$
$$period(s) = ((s - 1) \div 2n^{m-1}) + 1 \qquad (3.3)$$

Let $H = \{1, .., n\}$ the set of indices to the hue values chosen. Then, the hue value of a given stripe is set as follows

$$Colour : S \longrightarrow H$$
$$Colour(s) = \begin{cases} period(s) & Int(s) = medium \\ db\left[\frac{((s-1) \bmod 2n^2)+1}{2}\right] & otherwise \end{cases} \qquad (3.4)$$

where db is an ordered vector containing a De Bruijn sequence with window property of 2 like the following one [Salvi *et al.*, 2004]:

$$db = [1, 1, 2, 1, 3, 1, 4, 2, 2, 3, 2, 4, 3, 3, 4, 4] \qquad (3.5)$$

Every stripe with a certain hue value and a certain level of intensity can be labelled according to the set of labels $B = \{1, 2, 3, 4, 5, 6, 7, 8\}$ by using the following function

$$Label : H \times L \longrightarrow B$$
$$Label(h, l) = \begin{cases} h & l = high \\ h + n & l = medium \end{cases} \qquad (3.6)$$

The sequence of stripes generated by the proposed coding strategy and the De Bruijn sequence in Equation 3.5 can be represented using the labelling function in Equation 3.6 as follows

$$
\begin{aligned}
P_1 &= [5, 1, 5, 1, 5, 2, 5, 1, 5, 3, 5, 1, 5, 4, 5, 2, 5, 2, 5, 3, 5, 2, 5, 4, 5, 3, 5, 3, 5, 4, 5, 4] \\
P_2 &= [6, 1, 6, 1, 6, 2, 6, 1, 6, 3, 6, 1, 6, 4, 6, 2, 6, 2, 6, 3, 6, 2, 6, 4, 6, 3, 6, 3, 6, 4, 6, 4] \\
P_3 &= [7, 1, 7, 1, 7, 2, 7, 1, 7, 3, 7, 1, 7, 4, 7, 2, 7, 2, 7, 3, 7, 2, 7, 4, 7, 3, 7, 3, 7, 4, 7, 4] \\
P_4 &= [8, 1, 8, 1, 8, 2, 8, 1, 8, 3, 8, 1, 8, 4, 8, 2, 8, 2, 8, 3, 8, 2, 8, 4, 8, 3, 8, 3, 8, 4, 8, 4]
\end{aligned}
$$
$$(3.7)$$

where P_i is the sequence of stripes corresponding to period i. Note that the elements $\{1, 2, 3, 4\}$ correspond to high intensity stripes and 1 indicates the first hue value, 2 the second, etc. On the other hand, the elements $\{5, 6, 7, 8\}$ correspond to medium intensity stripes so that, in this case, 5 indicates the first hue value, 6 the second, etc. Note that in every period, all the medium intensity stripes share the same hue value, while the hue value of the high intensity stripes is directly chosen according to the De Bruijn sequence in Equation 3.5.

Figure 3.1a shows a pattern encoded according to the above strategy. The $n = 4$ hue values used are equally spaced in the Hue space so that $H = \{0°, 90°, 180°, 270°\}$.

Finally, the two intensity levels used are $L = \{0.35,\ 0.5\}$. The intensity profile of the pattern is shown in Figure 3.1b.

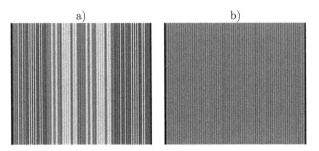

Figure 3.1: An example of new pattern. a) Hybrid pattern in the RGB space. b) Hybrid pattern intensity profile.

Note that 128 stripes have been coloured by using only 4 levels of hue. Indeed, as less colours are projected, higher immunity against noise is obtained. The proposed pattern allows two adjacent stripes to share the same hue value but not the same intensity level.

The aim of this pattern is to allow both intensity peaks (corresponding to the stripes central point) and edges between stripes to be detected.

Next section presents the algorithmic details concerning the detection of both intensity peaks and edges and the decoding of the new hybrid pattern.

3.3   Pattern segmentation and decoding

Given the system configuration of our experimental setup, i.e. a camera and a projector positioned aside and the pattern consisting of vertical stripes, the decoding process is based on horizontal scan-lines. In case that the projector and the camera are not approximately aside, a stereo pair rectification algorithm can be used to transform the images taking into account the geometry of the system [Fusiello *et al.*, 2000].

According to the aim of the proposed hybrid pattern, for every scan-line, the intensity peaks corresponding to the centre of every stripe and the edges between the stripes must be located. Afterwards, the detected stripes centres and the edges between adjacent stripes must be matched with the projected pattern in order to obtain correspondences.

3.3.1   Scan-line segmentation

First of all, we take advantage of the pattern square intensity profile in order to segment regions in the scan-line corresponding to medium intensity and high intensity stripes of the pattern. We define the M channel of a scan-line as a function of the RGB channels as follows

$$M(i) = \max(R(i), G(i), B(i)) \qquad (3.8)$$
$$i = 1..\text{width(scan-line)}$$

The second derivative of the M channel is used to distinguish between regions of maxima and minima intensity. In order to increase the immunity against noise the following numeric filtered linear derivative is used [Blais and Rioux, 1986]

$$g(i) = \sum_{c=1}^{o/2} (f(i+c) - f(i-c)) \qquad (3.9)$$

where f is the original function (in our case, the M channel of the current scan-line), g is the corresponding filtered linear derivative of order o, and i indicates the element index which is being filtered. This filtering operator is applied twice in order to obtain the filtered second derivative of the M channel. The order o of the filter must be chosen according to the apparent stripes width of the pattern in the images, which must be always greater than o (in our experiments we have used $o = 6$). The effects of applying this filtered derivative in a signal showing a maximum and a minimum is shown in Figure 3.2a. Note that the second derivative strongly enhances the intensity differences, producing more enhanced peaks. Then, the second derivative can be used to segment maximum and minimum intensity regions of the pattern by simply binarising it. Concretely, regions where the second derivative is less than 0 are binarised to 1 and inversely. In Figure 3.2b a portion of a scan-line of the M channel of a real image is shown. The square signal is the binarised second derivative of the scan-line. The results of segmenting the stripes by using this technique is also shown in Figure 3.2c-d. In this case, the M channel of an image with a human hand under the pattern illumination is shown, see Figure 3.2c. As can be seen in the resulting binarised image in Figure 2d, the second derivative is able to segment the stripes even in regions with low intensity (see the black background behind the hand). In order to avoid to process the image background a minimum intensity threshold applied to the M channel can be used in order to detect such regions. Note that the stripes corresponding to the human hand have all been correctly segmented even if there are contrast variations through the image.

The segmentation of the scan-line by using the second derivative allows us to distinguish between regions corresponding to medium and high intensity stripes. The centre of the stripe corresponding to every segmented region can be detected with sub-pixel accuracy using different peak detectors [Trucco et al., 1998]. After some experiments, we have found that given the width of the imaged stripes the peak detector obtaining better accuracy is the one based on a normalised centroid. This peak detector first normalises the pixel intensities by dividing it by the maximum intensity in the region. Then, it calculates the sub-pixel centre of mass of the region taking into account only those pixels for which its normalised intensity is higher than a certain threshold (set to 0.9 during the experiments).

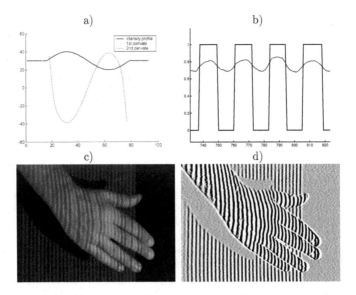

Figure 3.2: Stripes segmentation through a filtered second derivative. a) Behaviour of the 1st and 2nd derivative on a synthetic signal. b) Segmentation of maxima and minima using the 2nd derivative. c) M channel of an image of a human hand under the pattern illumination. d) Stripes segmented by using the 2nd Derivative.

The edges between adjacent stripes can be located by using the following strategy. Given the stripe centre of a high intensity stripe, the sub-pixel position of the surrounding edges corresponds to the two closest maxima in the function

$$g = dR^2 + dG^2 + dB^2 + dM^2 \qquad (3.10)$$

where dR, dG, dB and dM denotes the first derivative (calculated by the linear filter in Equation 3.9) of the Red, Green, Blue and M channel of the scan-line. This strategy is similar to the used in the edge-based stripe pattern by Zhang et al. [Zhang *et al.*, 2002] but in this case we take profit also of the alternating intensity profile of the pattern. The sub-pixel position of the maxima in g are calculated by using the peak detector by Blais and Rioux [Blais and Rioux, 1986] when a zero-crossing is detected in the first derivative of g.

3.3.2 Scan-line decoding

Once the stripes centres and the edges have been located, it is necessary to match them with the projected pattern, a process which is known as *pattern decoding*. The decoding strategy of our technique is based on the hue value of every segmented stripe in the scan-line.

The hue value of each segmented stripe is set to be the median hue value of all the pixels included in the corresponding region. Since only four hue values are projected, the hue of each region can be easily identified by using simple rules over the RGB colour like the following ones (taking into account the hue values used in the pattern proposed in Section 3.2)

$$\begin{cases} \max(R,G,B) = R \text{ and } G < 0.5R \text{ and } B < 0.5R & \Rightarrow \quad H = 0° \\ \max(R,G,B) = G \text{ and } B < 0.5G & \Rightarrow \quad H = 90° \\ \min(R,G,B) = R & \Rightarrow \quad H = 180° \\ \min(R,G,B) = G & \Rightarrow \quad H = 270° \end{cases} \qquad (3.11)$$

Then, for every scan-line a sequence of stripes is recovered and a hue value and an intensity level (medium or high) is assigned to each one of them. By using the labelling function presented in Equation 3.6 a numeric sequence representing the scan-line can be obtained. Ideally, if all the pattern stripes are segmented in the scan-line and their hue value and intensity level are correctly identified, the numeric sequence corresponding to the scan-line is equal to the encoded sequence in Equation 3.7. In such an ideal case, the matching of correspondences between the projected sequence and the perceived one is straightforward. However, in most cases either the whole sequence of projected stripes is not visible in the image scan-line or some of them are incorrectly labelled or disorders may occur.

In order to robustly solve the correspondence problem that arises we use dynamic programming as in [Cox *et al.*, 1996; Zhang *et al.*, 2002].

3.4 Improving the performance of coloured stripe patterns

In this section we present two modelling steps which should be always considered when working with coloured stripe patterns. Indeed, imperfections on the devices of the structured light system setup can affect the performance and accuracy of the technique. In the following sections two of these imperfections are introduced and a solution for minimising their negative effects are proposed.

3.4.1 RGB channel alignment

Ideally, colour cameras should perceive an intensity peak of white light at the same image coordinates in the three RGB channels. In practice there is an offset between the sub-pixel location of the intensity peaks in every RGB channel. This phenomenon is known as RGB channel misalignment. It is caused by spatial misalignments in the different CCD cells perceiving the red, green and blue light respectively. Although the order of these misalignments is usually below or around one pixel, it can produce higher order errors in 3D reconstruction. Furthermore, it can be easily shown experimentally that it is not only cameras the ones suffering from RGB channel misalignment, but also LCD projectors.

Some authors propose to reduce the camera RGB channel misalignment by viewing an object providing reference points (like a checkerboard) and locating such points in the three channels separately. Afterwards, an homography can be calculated relating the position of the points in the red channel with respect to the ones in the green channel, and another homography doing the same between the points in the blue and the green channel. These homographies are then used to reduce the misalignment in the images [Zhang *et al.*, 2002]. Nevertheless, this method totally ignores the RGB misalignment in the LCD projector.

We propose to minimise the RGB misalignment observed in the camera images taking into account both the camera and the projector at the same time. Since the decoding process is made through horizontal scan-lines, we propose to minimise the RGB channel misalignment in every scan-line. The algorithm is simple. A flat white panel is set in front of the camera and the projector at the typical working distance. Three patterns consisting of a sequence of narrow stripes separated by black gaps are projected sequentially: each one having red stripes, green stripes and blue stripes, respectively. Images of every projected pattern are taken by the camera. For every scan-line the sub-pixel position of every intensity peak is located with the detector by Blais and Rioux [Blais and Rioux, 1986]. In the image containing the red stripes, the red channel is used to locate the peaks. Similarly, the green channel is used in the image containing green stripes and the blue channel for the case of blue stripes. For every scan-line, the median of the sub-pixel offsets and the relative positions between the three channels are stored. We have observed that the relative positions of the channels coincide in all the scan-lines and that the relative offsets are very similar. That is why we finally store two unique offsets between the central channel and the other two. In our experimental setup we have found that in the images the central channel is the blue while the green channel is approximately at 1.26 pixels at its left, and the red channel is about 0.53 pixels at its right. These offsets have been named $^b\mathbf{H}_g$ and $^b\mathbf{H}_r$, respectively. In order to reduce the global misalignment observed in an image it is necessary to apply the offset $^b\mathbf{H}_g$ to the green channel and the offset $^g\mathbf{H}_r$ to the red one and then combine the transformed channels with the original blue channel in order to obtained the rectified image. Note that the intensity of every transformed pixel in the new channels must be interpolated from the neighbouring pixels in the corresponding source channel since the offsets are at sub-pixel precision.

3.4.2   Colour selection and calibration

The proposed hybrid pattern is based on the assumption that two different intensity levels will be distinguished in the image regions containing the projected pattern. This is necessary in order to be able to segment the stripes and to identify them as medium or high intensity stripes. Ideally, any discrete RGB instruction $\mathbf{c}$ with the same level of intensity i should produce the projection of light with the same intensity I so that the only varying parameter is the wavelength λ. Similarly, a perfect camera should be able to digitise any incident light of wavelength λ and a certain intensity I to a discrete RGB triplet $\mathbf{C}$ with intensity i. In real conditions, however, the mapping from the RGB projection instruction $\mathbf{c}$ to the imaged RGB triplet $\mathbf{C}$ is a strongly non-linear process. Several radiometric models of a structured light system composed of a LCD projector and a colour camera can be found in the bibliography, as for example, in [Caspi *et al.*, 1998; Nayar *et al.*, 2003]. Such models take also into account the reflective properties of the illuminated scene so that a model of the object albedo is also estimated. Indeed, the complete radiometric model identification of our experimental setup could be performed. However, taking into account that only 4 colours (hue values) and 2 different intensity levels are projected in the hybrid pattern, a simpler algorithm can be performed. Furthermore, since a one-shot technique is based on a unique pattern projection, the albedo of the object cannot be recovered so that this part of the radiometric model is unnecessary.

In Figure 3.3 the system response when projecting the four hue values with different intensities, from the point of view of the RGB instruction $\mathbf{c}$, is plotted. The response is the value of the M channel obtained from the camera image. These curves have been obtained by projecting solid patterns with each one of the 4 hue values with different intensity levels. A neutral colour panel has been used in order to not excessively perturb the projected light. As can be seen, the system response for each one of the colours is different and it is clearly non-linear. Note that in order to obtain the same value in the M channel of the image, each one of the 4 colours must be projected with a different intensity level. Therefore, as shown in Figure 3.3, by choosing the two desired values of the M channel, the required projecting intensities for each one of the colours can be approximately obtained using the curves. With this simple calibration procedure, the hybrid pattern can be adapted in order to produce a suitable system response without need of performing a whole radiometric calibration.

Another typical problem of a structured light system setup is the projector-camera colour crosstalk [Caspi *et al.*, 1998; Nayar *et al.*, 2003; Zhang *et al.*, 2002]. In order to

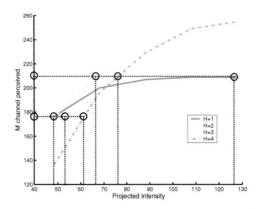

Figure 3.3: Non-linear response of the system when projecting the four selected hue values with different intensity levels.

see the role played by this phenomenon let us remember the model presented by Caspi et al. [Caspi *et al.*, 1998]

$$
\underbrace{\left(\begin{array}{c} R \\ G \\ B \end{array}\right)}_{\mathbf{C}} = \underbrace{\left(\begin{array}{ccc} a_{rr} & a_{rg} & a_{rb} \\ a_{gr} & a_{gg} & a_{gb} \\ a_{br} & a_{bg} & a_{bb} \end{array}\right)}_{\mathbf{A}} \mathbf{KP} \underbrace{\left\{\begin{array}{c} r \\ g \\ b \end{array}\right\}}_{\mathbf{c}} + \underbrace{\left(\begin{array}{c} R_0 \\ G_0 \\ B_0 \end{array}\right)}_{\mathbf{C_0}} \tag{3.12}
$$

where **c** is the RGB projection instruction sent to the projector and **C** the corresponding RGB triplet digitised by the camera. The consign **c** is actually modified by the non-linear behaviour of the projector which actually projects a colour denoted by **P**. **K** is a 3×3 matrix modelling the albedo of the illuminated object, and **A** is the colour crosstalk matrix, while **C₀** is the RGB tripled digitised by the camera when there is only ambient lighting. Therefore, **A** expresses how the RGB channels of the camera are affected by the RGB channels of the projector.

Experimentally it can be observed that usually the strongest crosstalk appears when projecting green, since it is not only detected by the green channel of the camera but also by the red one. In order to minimise the colour crosstalk, we perform two calibration processes. First, a solid pattern is projected using each one of the 4 calibrated colours for medium intensity stripes. These colours are defined by the 4 hues selected in Section 3.2 and the 4 intensities selected in this section which aim to produce the same medium intensity in the M channel. Then, the following linear system is defined

$$\left(\begin{array}{cccc} R_1 & R_2 & R_3 & R_4 \\ G_1 & G_2 & G_3 & G_4 \\ B_1 & B_2 & B_3 & B_4 \end{array}\right) = \mathbf{A}_m \left(\begin{array}{cccc} r_1 & r_2 & r_3 & r_4 \\ g_1 & g_2 & g_3 & g_4 \\ b_1 & b_2 & b_3 & b_4 \end{array}\right) \tag{3.13}$$

where (R_i, G_i, B_i) is the average colour perceived by the camera when the instruction (r_i, g_i, b_i) is sent to the projector and using a colour neutral panel ($\mathbf{K} = \mathbb{I}_3$). Matrix $\mathbf{A}_m$ can be numerically calculated by using singular value decomposition. The inverse of this matrix can be used to partially remove the colour crosstalk of the stripes identified during the segmentation process as medium intensity stripes. Similarly, a matrix $\mathbf{A}_h$ is calculated by projecting the colours corresponding to the high intensity stripes.

3.5 Experimental results

In this section we show some experimental results validating the proposed hybrid pattern. The performance of the new pattern is evaluated by calibrating the devices of the structured light system and triangulating the correspondences.

In this case, the experimental setup consists of an Olympus Camedia digital camera and a Mitsubishi XL1U LCD projector which are positioned aside with a relative direction angle of about 15°. Both devices operate at 1024×768 pixels. The measuring volume is about 30 cm high, 40 cm wide and 20 cm deep.

The calibration of the system has been performed according to the steps presented in the previous section and summarised in Figure 3.4. The figure also shows the one-shot shape acquisition process and its interaction with the data obtained during the system calibration. Note that the image rectification step is not really necessary if the camera and the projector are positioned aside with similar tilt angle.

During the experiments three different patterns have been tested which are now briefly presented.

Edge-based pattern: the stripe pattern proposed by Zhang et al. [Zhang et al., 2002] has been chosen as sample of pattern where it is possible to match points belonging to the edges between adjacent stripes. The pattern is composed of 125 vertical stripes 7 pixels wide coloured using 8 different hue levels and is shown in Figure 3.5a. The edges are located in each scan-line by searching local maxima of the following function

$$f = dR^2 + dG^2 + dB^2 \tag{3.14}$$

where dR, dG and dB are the intensity gradients of the red, green and blue channels on the current scan-line. Every edge is labelled as $(dR(e_i), dG(e_i), dB(e_i))$ where e_i is the position of the edge on the scan-line. The pattern is generated so that every combination of three consecutive edge labels appears only once as maximum (window property equal to 3). The correspondence problem between the sequence of edges located in a scan-line

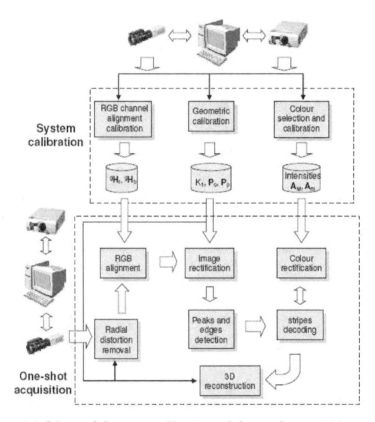

Figure 3.4: Schema of the system calibration and the one-shot acquisition steps.

and the sequence of emitted edges is solved by using dynamic programming [Zhang *et al.*, 2002].

Peak-based pattern: a pattern with 64 coloured slits of 7 pixels separated by black gaps of the same width. Four colours have been used to generate the pattern which is shown in Figure 3.5b. The sequence of coloured stripes has window property equal to 3. Similar patterns can be found in the bibliography like, for example, the one by Chen et al. [Chen *et al.*, 1997] or the one by Monks et al. [Monks *et al.*, 1992]. In order to locate the slits in a camera scan-line maxima in the M channel are used. Every maximum corresponds approximately to the bounds of a coloured slit. A coloured slit is said to be between two consecutive maxima if the gradient of M in this region is predominantly descending. The accurate position of the central point of a coloured slit is located by using a normalised centroid peak detector [Trucco *et al.*, 1998]. Then, since only 4 colours are projected, the colour of the detected slits can be easily segmented. The matching of the detected slits on the current scan-line and the sequence of projected ones is also made through dynamic programming based on the RGB components of the slits.

Hybrid pattern: the pattern containing 128 stripes generated according to the new coding strategy which was already presented in Figure 3.1. Both medium and high intensity stripes are 7 pixels wide.

Figure 3.5: One-shot patterns used for comparing the performance of the hybrid pattern. a) Stripe pattern by Zhang et al. b) Multi-slit pattern similar to the one proposed by Monks et al.

3.5.1 Quantitative results

The selected techniques have been first compared in terms of accuracy and resolution. A measure of the accuracy has been obtained by reconstructing a plane positioned in front of the camera and the projector at a distance of about 1.2 m. The accuracy has been characterised by the mean and the standard deviation of the distances between the reconstructed $3D$ points and the fitted plane. The figure for evaluating the resolution is the number of $3D$ reconstructed points which is the number of correspondences found.

When using the peak-based pattern, the $3D$ points are obtained by reconstructing the intensity peaks detected in the image. In the case of the edge-based pattern, the detected edges are used to triangulate $3D$ points. On the other hand, the hybrid pattern allows to obtain $3D$ points by reconstructing both intensity peaks and edges. Let us introduce the following notation to refer to the basic reconstruction strategies allowed in the hybrid pattern:

- M strategy: reconstruction of intensity peaks (maxima) corresponding to the centre of high intensity stripes.

- m strategy: reconstruction of intensity peaks (minima) corresponding to the centre of medium intensity stripes.

- E strategy: reconstruction of edges between adjacent stripes.

In fact, not only the individual strategies M, m and E have been used when testing the hybrid pattern, but also some combinations of them, namely $M + m$, $E + M$ and $E + M + m$.

Table 3.1 shows the numerical results obtained by every pattern and the different reconstruction strategies allowed by each one. In terms of accuracy the best result is obtained by the peak-based pattern. In second position we find the hybrid pattern when using the M strategy. In both cases, the average and the standard deviation of the error is much more lower that when using the edge-based pattern. Furthermore, we note that the accuracy of the hybrid pattern when reconstructing only edges (E strategy) is also better that the accuracy of the edge-based pattern. When using the m strategy with the hybrid pattern, the average error increases considerably while the standard deviation remains quite similar. This points out that the reconstruction of minima is much more affected by noise. Therefore, finding the sub-pixel position of an intensity minimum is a much more sensitive and inaccurate process than locating an intensity maximum. This is probably caused because the signal-to-noise ratio is lower in the medium intensity stripes of the hybrid pattern. In terms of resolution we note that strategies based on intensity peaks (M and m) obtain about half of the resolution that when reconstructing edges (E strategy). Therefore, from the numerical results it is not easy to decide which is the best individual strategy because there is a trade-off between accuracy and resolution.

Let us now analyse the results when combining different reconstruction strategies thanks to the hybrid pattern. Using more than one strategy allows the resolution to be largely increased, see last column of Table 1. In general, the combination $M + m$ obtains the same order of resolution that the E strategy. Note however, that the accuracy of the hybrid pattern when using $M + m$ is slightly better than using the hybrid pattern and the E strategy. With respect to the edge-based pattern, the improvement on accuracy is much more significant. On the other hand, we can emphasise that the combination $E + M$ gets a resolution about 1.5 times greater than when using E or $M + m$ and obtains similar accuracy. Finally, the combination $E + M + m$ doubles the resolution of the E strategy. However, we observe a worse accuracy (the standard deviation of the error is higher). It seems therefore, that the inclusion of the minima intensity peaks degrades the performance of the hybrid pattern in terms of accuracy.

Table 3.1: Accuracy results of plane fitting obtained from the three strategies. E denotes edges, M maxima intensity peaks and m minima intensity peaks.

Pattern type	Strategy	Mean (mm)	StDev (mm)	Reconstructed points
peak-based	M	0.30	0.22	1964
edge-based	E	0.63	0.37	3938
hybrid	E	0.51	0.34	3886
hybrid	M	0.43	0.33	1943
hybrid	m	0.52	0.41	1943
hybrid	M+m	0.43	0.34	3886
hybrid	E+M	0.48	0.34	5829
hybrid	E+M+m	0.46	0.44	7973

The results of the patterns are qualitative compared in the following section. These results will help us to decide which combination of reconstruction strategies give better results when using the proposed hybrid pattern.

3.5.2 Qualitative results

In this section we present and discuss the visual appearance of several reconstructions obtained with the selected patterns and the different reconstruction strategies. First of all we analyse the plane reconstruction explained in the previous section. In Figure 3.6a the reconstruction obtained using the peak-based pattern is shown. Note that the reconstructed surface is very smooth confirming the accuracy results presented before. Figure 3.6b shows the surface obtained when using the edge-based pattern. Note that some visual artifacts, concretely vertical ridges, appear along the surface. Figure 3.6c-d shows the surfaces obtained with the hybrid pattern when reconstructing only maxima intensity peaks and edges, respectively. Note that in this case a smooth surface is also obtained with the M strategy, while some ridges also appear when using the E strategy. Nevertheless, we remark that the ridges obtained with the hybrid pattern when reconstructing edges are much smoother that the ones appearing in the reconstruction provided by the edge-based pattern. This fact seems to confirm the accuracy results obtained by both techniques. The reconstruction obtained by the hybrid pattern when using the $M + m$ strategy is shown in Figure 3.6e. Note that a smooth plane is still obtained. The result given by the hybrid pattern and the $E + M$ strategy is depicted in Figure 3.6f. Even if a small loss of accuracy has been predicted in the previous section, we can observe that the visual appearance is still better that when using the edge-based pattern. Nevertheless, as expected, the inclusion of the edges in the reconstruction process make some smooth ridges to appear. Finally, Figure 3.6g presents the reconstruction when using the strategy $E + M + m$. The appearance is bit noisier than the obtained by the $E + M$ strategy.

Figure 3.7 presents three colour neutral test objects, namely a mask, a horse statue and a sun statue. The images corresponding to the projected patterns (peak-based, edge-based and hybrid pattern) are also shown. The reconstruction results are plotted in Figure 3.8.

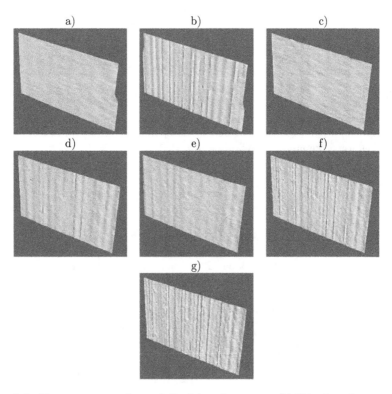

Figure 3.6: Plane reconstruction. a) Peak-based pattern. b) Edge-based pattern. c) Hybrid pattern: M strategy. d) Hybrid pattern: E strategy. e) Hybrid pattern: M+m strategy. f) Hybrid pattern: E+M strategy. g) E+M+m strategy.

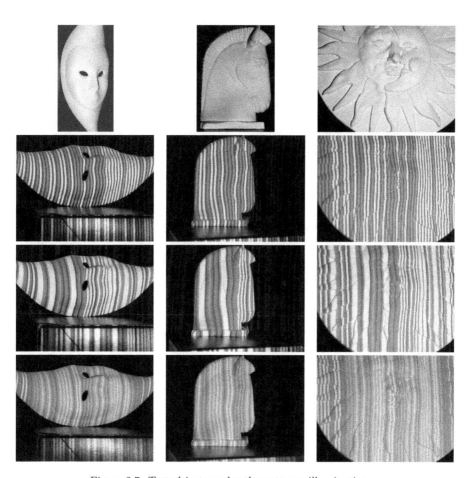

Figure 3.7: Test objects under the patterns illumination.

As expected, the peak-based pattern (first row) and the hybrid pattern using the $M+m$ strategy (second row) obtain both smooth surfaces and almost absence of artifacts. However, the reconstructions obtained by the peak-based pattern are too much smoothed due to the low resolution achieved, so that details like the mask mouth and the horse eye are not visible. Note that in the result corresponding to the hybrid pattern and the $M+m$ strategy such details begin to be appreciated. The gain in visual appearance obtained by the hybrid pattern and the $M+m$ strategy is confirmed in the sun reconstruction, where the eyes, nose and mouth are better represented than in the case of the peak-based pattern.

The edge-based pattern (third row) shows pretty performance producing highly detailed surfaces with absence of important holes. On the other hand, vertical ridges appear, degrading the smoothness of the surfaces. If we compare the edge-based pattern and the hybrid pattern using the $M+m$ strategy, the former seems to get a bit more level of detail. However, since the latter does not have the inconvenience of the vertical ridges, it is difficult to decide which results are better from a qualitative point of view.

Finally, the last two rows of Figure 3.8 present the results obtained by the hybrid pattern when using the $E+M$ and $E+M+m$ strategies. The $E+M$ strategy obtains a higher level of detail (see for example the horse mouth, which was not detected in the previous cases) than the previous techniques. Due to the inclusion of the edges in the triangulation process, the vertical ridges also appear. However, these artifacts are smoother than the ones observed in the edge-based pattern results. On the other hand, it seems that some additional small holes appear and that some contours near shadows are less well defined. Note also that the differences between the results of the $E+M$ strategy and the $E+M+m$ strategy can be hardly distinguished.

In general, we think that the results obtained by the hybrid pattern are globally better than the ones obtained by the edge-based pattern proposed by Zhang et al. [Zhang *et al.*, 2002]. We think that the increase in resolution achieved by the combination of edges and intensity peaks allows some small details of the objects to be better reconstructed. In order to show this, Figure 3.9 presents an ampliation of the sun reconstruction focusing on the zone corresponding to the nose, eyes and mouth. As can be seen in Figure 3.9b, the reconstruction obtained by the peak-based pattern is quite poor. Note that the quality of the reconstruction is already improved when using the hybrid pattern and the $M+m$ strategy as shown in Figure 3.9e. Even if the reconstruction obtained by the edge-based pattern (see Figure 3.9c) shows great level of detail, we can see that the results obtained by the hybrid pattern and the $E+M$ and $E+M+m$ strategies (Figure 3.9d and f) are even better. Note that the shape of the sun nose and mouth and the nose corresponding to the moon are much more clear in the hybrid pattern reconstructions. We can also point out that the fold appearing in the left cheek of the sun is better appreciated in the hybrid pattern results.

Finally, we present to reconstructions obtained with the hybrid pattern taking into account slightly coloured objects. One-shot techniques projecting colourful patterns have usually problems when dealing with non-colour neutral objects. We first show the results obtained when reconstructing a human hand. The skin usually introduces a strong gain on the red component of the pattern and at the same time attenuates the whole luminosity reaching the camera. Figure 3.10a shows the picture of the hand with the projected

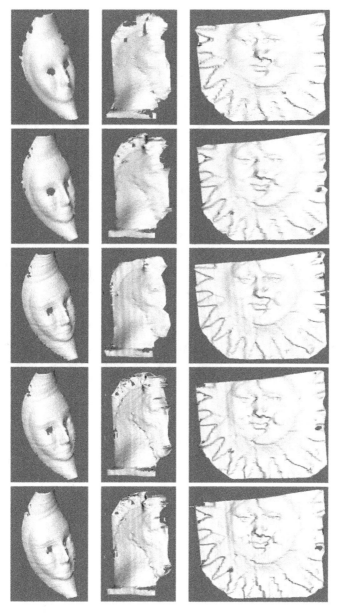

Figure 3.8: Reconstruction results of the test objects. The first row corresponds to the peak-based pattern; second row: hybrid pattern and the $M + m$ strategy; third row: edge-based; fourth row: hybrid pattern and the $E + M$ strategy; sixth row: hybrid pattern and $E + M + m$ strategy.

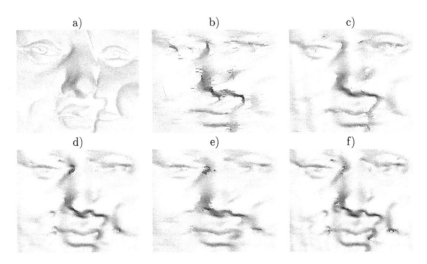

Figure 3.9: Zone enlargement of the sun reconstruction. a) Picture of the real object. b) Peak-based pattern. c) Edge-based pattern. d) Hybrid pattern: M+E strategy. e) Hybrid pattern: M+m strategy. f) Hybrid pattern: E+M+m strategy.

pattern. The stripe and colour segmentation obtained is shown in Figure 3.10b. The reconstruction results are plotted in form of points and surface in Figure 3.10c-d. As can be seen, the level of detail is as expected very high (see the veins passing through the hand) taking into account that a unique pattern is projected. The second example consists of three coloured sheets of papers as shown in Figure 3.10e. The obtained reconstruction is presented in Figure 3.10f.

3.6   Conclusions

This chapter has presented a new coloured pattern which is able to provide a large number of correspondences in a single shot. The new pattern aims to improve similar existing patterns in terms of resolution and accuracy. Among the one-shot colour-based techniques two of the most frequent patterns are based on multi-slits or stripes. In multi-slit patterns the intensity profile has periodic peaks corresponding to the central point of the slits which are used to find correspondences. In stripe patterns, since the intensity profile is flat, edges between adjacent stripes are used to find correspondences. Both types of patterns are usually coloured according to De Bruijn sequences. The advantage of stripe patterns over multi-slit patterns is that the resolution is higher since no black gaps must be inserted. However, the number of hue values required is also higher since adjacent stripes must be different.

In this chapter a new hybrid pattern has been proposed which combines the properties

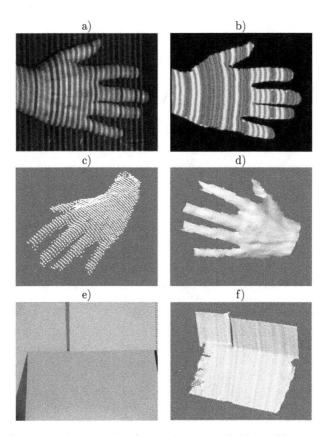

Figure 3.10: Reconstruction example of non-colour neutral objects. Human hand: a) Skin under the pattern illumination. b) Colour segmentation of the received pattern. c) Cloud of reconstructed points. d) Surface fitting the reconstructed hand. Coloured papers: e) the coloured sheets used in the experiment. f) The reconstructed surface.

of a multi-slit pattern and the ones of an edge-based pattern. First of all, the new coding strategy allows a stripe pattern to be defined by using less hue values than usual. The new pattern contains $2n^m$ stripes, where the window length m is set to be 3, and n, the number of different hue values used, remains free in order to obtain the desired resolution. We remind that classical multi-slit patterns contain n^m slits, since consecutive slits can have the same hue. In the case of stripe patterns, the number of coloured bands is $n(n-1)^{m-1}$ [Hügli and Maître, 1989]. Therefore, the resolution of the new hybrid pattern given a certain number of hue values is always larger than in classic multi-slit and stripe patterns.

The hybrid pattern has a multi-slit structure in the Intensity channel while it has a stripe appearance in the RGB space. Then, the odd stripes have a medium-intensity value while the even stripes have high-intensity. Since all the stripes are coloured, both maxima and minima intensity peaks can be located in the images. Furthermore, edges between stripes can also be detected with accuracy. This allows the pattern resolution to be increased in a factor up to 2.

The new pattern has been compared to an edge-based pattern and a peak-based pattern. The performance of the three patterns have been analysed in a shape acquisition framework by calibrating the camera and the projector. The accuracy and resolution of each technique has been characterised by reconstructing a plane and measuring the average and the standard deviation of the fitting error, and the number of $3D$ points, respectively. A first conclusion is that the accuracy of the sub-pixel correspondences based on intensity peaks is higher than when matching edges as shown by the reconstruction results. Indeed, the sub-pixel estimation of an intensity peak is more stable than the sub-pixel location of an edge between coloured stripes. This numerical results are confirmed by the visual appearance of the reconstructions. When reconstructing intensity peaks, the obtained surfaces are smoother, while in the case of edges, visual artifacts like ridges appear.

The peak-based pattern has obtained the better results in terms of accuracy. Nevertheless, the number of correspondences provided by this pattern is quite low. Concretely, it is about half of the obtained by the edge-based pattern. The better results obtained with the hybrid pattern arise when reconstructing only the maxima intensity peaks. In this case, however, the resolution is also half of the obtained by the edge-based pattern. The accuracy of the hybrid pattern when reconstructing only edges is still better than the one obtained by the edge-based pattern. This fact is visually confirmed by observing the plane reconstruction. The numerical results show that the accuracy diminishes when using the hybrid pattern and reconstructing only minima intensity peaks. This confirms that the medium-intensity stripes are more sensitive to noise. However, the most interesting results are obtained when using the hybrid pattern and combining edges and maxima intensity peaks. In this case, the accuracy is not so good as for the peak-based pattern, but it is better than the edge-based pattern. In addition to this, the resolution is increased in a factor of 1.5. The resolution of the edge-based pattern is overcome by a factor of 2 when using the hybrid pattern and when reconstructing edges and maxima and minima intensity peaks. Nevertheless, the accuracy is slightly worse than when reconstructing only edges and maxima intensity peaks.

Several objects containing high level of details have been reconstructed by using the three selected patterns. The visual appearance of the reconstructions confirm that when using only intensity peaks the obtained surfaces are smoother. Otherwise, when using edges periodic ridges appear. On the other hand, the importance of increasing the resolution is visually confirmed. In fact, while with the peak-based pattern coarse object's reconstructions are obtained, with the edge-based and the hybrid pattern they are more detailed. Furthermore, it has been shown that when using the hybrid pattern and reconstructing edges and maxima intensity peaks, more object's details are distinguished in the reconstructions. Finally, two results showing the performance of the new hybrid pattern when reconstructing more difficult textures, like human skin, have been presented, obtaining good results even if the colours of the pattern are perturbed by the object's albedo.

Chapter 4

An approach to visual servoing based on coded light

The aim of visual servoing is to control a robot by using visual information provided by a vision sensor. This chapter presents an overview of visual servoing techniques and focuses on the projection of structured light as a reliable way to provide visual features. An approach to visual servoing based on coded structured light is presented. The approach is validated through experimental results and the advantages and constraints are discussed.

4.1 Introduction

Visual servoing is a largely used technique which is able to control on-line robots by using data provided by visual sensors. Nowadays, the most typical sensors used in a visual servoing framework are CCD cameras. In this case, some features extracted from images are used as feedback in a control loop which leads to the execution of a robotic task like positioning with respect to static objects or target tracking. A comprehensive survey on the main visual servoing approaches can be found, for example in [Hutchinson *et al.*, 1996].

A necessary condition to apply visual servoing is that visual features must be available in the images. Therefore, positioning a robot with respect to a uniform or non-textured object, wall, obstacle, etc. is not feasible since no visual features can be extracted. In order to remove this limitation, structured light can be used by project visual features onto the object of interest. However, this solution can be insufficient. In visual servoing it is usual to define the goal robot position by the image perceived in such configuration, the so-called *desired image*. If the initial and desired robot positions are quite far, matching the visual features extracted from the initial and desired images can be very difficult. This happens, for example, when using points as visual features. A way to solve this problem is to project coded structured light patterns. As seen in Chapter 2, by including a coding strategy on the pattern, the visual features can be robustly identified and matched when viewed from different points of view.

The aim of this chapter is to propose an approach to visual servoing based on coded structured light projection by using a static projector. Using coded light patterns in a visual servoing framework is a research area which has not been yet investigated. Therefore, this chapter intends to make a first step contribution in this field showing the potentiality of such a new approach. The case of an onboard structured light emitter attached to the camera of the robot end-effector will be studied in Chapter 5.

The chapter is organised as follows. First of all, a brief review of the mathematical basis of visual servoing and an overview of the existing techniques are presented in Section 4.2 and Section 4.3, respectively. These reminders are here included in order to clarify the reading of the rest of the chapter as well as following chapters. Afterwards, Section 4.5 exposes a new approach to visual servoing based on coded structured light. The choice of the coded pattern as well as its segmentation and decoding are discussed. Some experimental results of the new approach are presented in Section 4.6. The chapter ends with conclusions and some perspectives.

4.2   The fundamentals of visual servoing

Two main configurations in visual servoing exist. In both cases, cameras are used to extract visual features from the images which will be used for controlling purposes. The most typical configuration consists in attaching a camera to the end-effector of the robot. Such configuration is known as *eye-in-hand* and it is represented in Figure 4.1a. In this case, the visual features extracted from the images correspond to some object of the environment. Another configuration known as *eye-to-hand* is shown in Figure 4.1b. In this case, the camera remains static observing the end-effector of the robot and its environment. The most usual configuration used in robotics is the eye-in-hand configuration which will be considered the default hereafter.

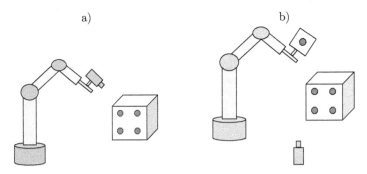

Figure 4.1: Typical configurations for visual servoing.

Visual servoing is based on the relationship between the variation of the robot pose and the consequent variation of the visual features extracted from the images. The robot pose is expressed as an element of the configuration space of rigid bodies noted as SE_3 (Special Euclidean group). Therefore, a robot moves in a three-dimensional space according to 6 degrees of freedom: three for displacements and three for rotations. The visual features are contained in a $\mathbb{R}^k$ vector of the form

$$\mathbf{s} = \mathbf{s}(\mathbf{r}(t)) \tag{4.1}$$

where $\mathbf{r}(t)$ is the relative pose between the camera and the environment at time t. Therefore, the time derivative of the visual features depends on the pose variation as pointed out by the well-known equation [Feddema *et al.*, 1991; Espiau *et al.*, 1992]

$$\dot{\mathbf{s}} = \frac{\partial \mathbf{s}}{\partial \mathbf{r}} \dot{\mathbf{r}} = \mathbf{L_s} \mathbf{v} \tag{4.2}$$

where $\mathbf{L_s}$ is the image jacobian so-called *interaction matrix*, and $\mathbf{v} = (V_x, V_y, V_z, \Omega_x, \Omega_y, \Omega_z)$ the camera velocity screw assuming a static environment.

4.2.1 Task function

A robotic task can be described by a function which must be regulated to 0 [Samson *et al.*, 1991; Espiau *et al.*, 1992]. In visual servoing the goal of the task is usually defined by the visual features corresponding to the desired camera-environment relative pose. These desired visual features are acquired during a learning stage by bringing the robot to the desired configuration. Concretely, the task function is noted as the following m−dimensional vector

$$\mathbf{e} = \mathbf{C}(\mathbf{s} - \mathbf{s}^*) \tag{4.3}$$

where $\mathbf{s}$ are the visual features corresponding to the current state and $\mathbf{s}^*$ denotes the visual features values in the desired state. $\mathbf{C}$ is a $m \times k$ combination matrix that must be of full rank $m \leq k$ in order to produce the m independent components of $\mathbf{e}$. The aim of visual servoing is to regulate the task function $\mathbf{e}$ to $\mathbf{0}$ so that $\mathbf{s} - \mathbf{s}^* = \mathbf{0}$.

The task $\mathbf{e}$ controls m degrees of freedom from a total of n. When $m < n$ it means that a virtual link of class N is fulfilled so that $m = n - N \leq k$.

A usual choice for the combination matrix $\mathbf{C}$ is [Espiau *et al.*, 1992]

$$\mathbf{C} = \mathbf{W}\widehat{\mathbf{L_s}}^+ \tag{4.4}$$

where $\widehat{\mathbf{L_s}}^+$ is the pseudoinverse of a model of the interaction matrix and $\mathbf{W}$ is an $m \times 6$ matrix of full rank m having the same kernel that $\mathbf{L_s}$. The choice of $\mathbf{W}$ depends on the number of visual features k. For example, let us consider the following cases

- $\text{rank}(\mathbf{L_s}) = m = k \Rightarrow \mathbf{W} = \widehat{\mathbf{L_s}}$. In this particular case $\mathbf{C} = \mathbf{I}_m$.

- rank$(\mathbf{L_s}) = m < k \Rightarrow$ the rows of $\mathbf{W}$ are the m vectors forming the base of the row space generated by $\widehat{\mathbf{L_s}}$.

4.2.2   Control law

A simple control law can be defined in order to fulfill the task $\mathbf{e}$. We assume that the combination matrix $\mathbf{C}$ is constant so that the derivative of the task function (4.3) is

$$\dot{\mathbf{e}} = \mathbf{C}\dot{\mathbf{s}} \tag{4.5}$$

and taking into account that

$$\dot{\mathbf{s}} = \mathbf{L_s}\mathbf{v} \tag{4.6}$$

we have that

$$\dot{\mathbf{e}} = \mathbf{CL_s}\mathbf{v} \tag{4.7}$$

then, by imposing an exponential decrease of the task function

$$\dot{\mathbf{e}} = -\lambda\mathbf{e} \tag{4.8}$$

being λ a positive gain, we find

$$-\lambda\mathbf{e} = \mathbf{CL_s}\mathbf{v} \tag{4.9}$$

from this expression a proportional control law can be built by using a model of the interaction matrix $\widehat{\mathbf{L_s}}$

$$\mathbf{v} = -\lambda(\widehat{\mathbf{CL_s}})^{-1}\mathbf{e} \tag{4.10}$$

which is equal to

$$\mathbf{v} = -\lambda(\widehat{\mathbf{CL_s}})^{-1}\mathbf{C}(\mathbf{s} - \mathbf{s}^*) \tag{4.11}$$

If it is not possible to estimate all the parameters of the interaction matrix at each iteration, a typical choice is to set $\widehat{\mathbf{L_s}}$ as the interaction matrix evaluated at the desired state noted as $\mathbf{L_s^*}$ or $\mathbf{L_s}(\mathbf{e}^*)$.

Then, from the control law in (4.10) and the expression in (4.7) the variation of $\mathbf{e}$ is defined by the following equation

$$\dot{\mathbf{e}} = -\lambda\mathbf{CL_s}(\widehat{\mathbf{CL_s}})^{-1}\mathbf{e} \tag{4.12}$$

so that a sufficient condition for ensuring the convergence to its desired state is

$$\mathbf{CL_s}(\widehat{\mathbf{CL_s}})^{-1} > 0 \tag{4.13}$$

4.3 Overview of visual servoing approaches

This section presents a brief overview of visual servoing techniques. A classic classification based on the type of visual features has been used similarly than in [Hutchinson *et al.*, 1996; Chaumette, 2002]. However, the techniques exploiting structured light have been classified apart in a new group.

4.3.1 Position-based visual servoing

In the case of position-based visual servoing, $3D$ information computed from the image(s) are used in the control law. In this case, the object model is known and its pose is estimated by using algorithms based on points [Horaud *et al.*, 1989; Dementhon and Davis, 1995; Haralick *et al.*, 1989; Yuan, 1989; Lowe, 1991; Wilson *et al.*, 1996], straight lines [Dhome *et al.*, 1989], polyhedric objects [Drummond and Cipolla, 2002], conics [De Ma, 1993] and some quadrics like spheres and cylinders [Dhome *et al.*, 1990].

A classical advantage granted to position-based approaches is their ability to produce good camera trajectories since the control is made in the cartesian space [Chaumette, 1998]. In addition to this, the rotational velocities can be decoupled from the translational ones, even if this only holds when the system is perfectly calibrated. There are three main drawbacks inherent to position-based approaches. First, the pose estimation or $3D$ reconstruction algorithms are sensible to image noise. Secondly, since no control is made in the image space, the features used for the reconstruction can get out of the image bounds. Finally, the stability analysis taking into account calibration errors is in most cases impossible to face since it depends on the algorithm of pose estimation.

4.3.2 Image-based visual servoing

Image-based or $2D$ visual servoing consists in using visual features directly calculated from the images which are used as input in the control scheme. Thus, this type of approaches tend to avoid the use of any object model.

Former works started using image points which are still today one of the most popular primitives [Feddema *et al.*, 1991; Espiau *et al.*, 1992; Bien *et al.*, 1993; Hager, 1997; Rives, 2000; Lots *et al.*, 2001; Collewet and Chaumette, 2002]. Other $2D$ primitives have been modelled like straight lines [Espiau *et al.*, 1992; Malis *et al.*, 2002], segments, circles, ellipsis, cylinders and spheres [Espiau *et al.*, 1992; Marchand and Chaumette, 1999].

On the recent years, more complicated primitives have been taken into account. For example, complex contours [Colombo and Allotta, 1999; Collewet and Chaumette, 2000], the principal components of the image [Deguchi, 1997] and image moments [Chaumette, 2004].

Some other works tend to combine different visual features cited above in a unique control scheme. For example, in [Corke and Hutchinson, 2001], point coordinates, the area enclosed by points and the angle between segments are used to improve the performance of the system.

Image-based visual servoing is traditionally robust against modelling errors of the system like camera calibration errors. Furthermore, since control is made in the image space, it is easier to design strategies to avoid image features going out the image bounds [Mezouar and Chaumette, 2002; Corke and Hutchinson, 2001]. However, since the features are usually strongly coupled, the generated camera trajectory cannot be very suitable. Some efforts in order to decouple degrees of freedom can be found for example in [Corke and Hutchinson, 2001; Mahony *et al.*, 2002; Tahri and Chaumette, 2004]. Other drawbacks are the possibility of reaching a singularity in the control law or falling into local minima [Chaumette, 1998].

4.3.3   Hybrid visual servoing

This approach combines $2D$ with $3D$ features. In case of knowing the model of the object, classic pose recovering algorithms can be used to estimate some $3D$ features as in position-based visual servoing. However, several model-free approaches have been presented. A lot of approaches have been done for the case when the desired image is known. Some of them are based on recovering the partial pose between the camera and the object from the desired and the current image [Malis *et al.*, 1999; Morel *et al.*, 2000; Lots *et al.*, 2000; Malis and Chaumette, 2002]. The obtained homography is decomposed in a rotation matrix and a scaled translation. Note that if both displacement components are directly used in a control law, a model-free position-based visual servoing scheme arises as in [Basri *et al.*, 1998]. However, the most usual choice is to combine part of the $3D$ information recovered with $2D$ features like an image reference point. Other approaches exist, like the one presented in [Schramm *et al.*, 2004], where the depth distribution of the object is explicitly included in the visual features. Another example is found in [Andreff *et al.*, 2002], where the plücker coordinates of $3D$ lines are used, so that the depth to the lines are estimated from a pose calculation algorithm assuming a partial knowledge of the object structure. When the desired image is unknown, the rotation to be executed can be calculated by doing a local reconstruction of the object normal in a certain point [Questa *et al.*, 1995; Colombo *et al.*, 1995; Sundareswaran *et al.*, 1996; Alhaj *et al.*, 2003; Collewet *et al.*, 2004].

Typical advantages of hybrid approaches are: they are usually model-free (do not require to know the object model even if in most cases the desired image must be known); they allow control in the image since $2D$ information is included; they can exhibit decoupling between translational and rotational degrees of freedom; stability analysis in front of camera and robot calibration errors is often feasible [Morel *et al.*, 2000; Malis and Chaumette, 2002]. On the other hand, the main drawback that can appear is the sensibility to image noise affecting the partial pose algorithm.

4.3.4   Dynamic visual servoing

The analysis of the $2D$ motion appearing in a sequence of images can be used to obtain geometric visual features which can be then used in a visual servoing scheme like the presented in the previous sections [Colombo *et al.*, 1995; Sundareswaran *et al.*, 1996; Collewet *et al.*, 2004; Alhaj *et al.*, 2003]. However, if the visual features are, for example, the parameters of the $2D$ motion model itself, a dynamic visual servoing scheme can be defined [Santos-Victor and Sandini, 1997; Crétual and Chaumette, 2001].

Basically, these techniques define the vision task in terms of dynamic visual features so that, for example, the system can be controlled in order to observe a desired $2D$ motion field along the sequence of acquired images.

4.4   Combining visual servoing and structured light

Although the large domain of applications that can be faced with classic visual servoing, there are still some open issues. As already mentioned, classic techniques cannot cope, for example, with the simple problem of keeping a mobile robot running parallel to a wall containing no landmarks or easily distinguishable visual features. This problem can be generalised to any task where it is necessary to position the camera with respect to an object with uniform appearance so that it is not possible to extract visual features like characteristic points, straight lines, contours, regions, etc. A possible solution to this problem is to use structured light emitters to project visual features in such objects. Thus, the application field of visual servoing can be enlarged to applications like painting, welding, trimming or navigation in general.

As already mentioned in Chapter 2 there are many robotic applications taking profit of structured light. In the following section we present an overview of applications in different fields of robotics. Afterwards, we focus on applications explicitly using image-based visual servoing and structured light.

4.4.1   Applications of structured light in robotics

Applications in mobile and autonomous robots

In this section some examples of vision tasks performed with the aid of structured light in mobile or underwater robots are reported.

Range sensing: robots can be equipped with a structured light source and a camera for measuring distances. In this case the range measures are obtained from the images by using triangulation based on the calibration of the camera and the light source. We must differentiate this range acquisition technique from the laser telemetry based on the time-of-flight of laser pulses which is also widely used in robotics [Nevado *et al.*, 2004; Matthies *et al.*, 2002]. In [Kondo and Tamaki, 2004] two laser pointers projecting one each a light spot are placed onboard an underwater robot for calculating distances to obstacles in front of the vehicle. A similar setup is used in [Sun *et al.*, 2004] which allows a glass climbing robot to calculate its orientation to the window-pane and the window frame. In [Sazbona *et al.*, 2005], a laser projects several segments so that the range is a function of the observed segments inclination. Hattori and Sato developed a small head composed of a camera and a diode laser and a rotating mirror which projects a fast sequence of binary encoded patterns for dense reconstruction of the environment.

Object recognition: a typical structured light sensor composed of a laser plane and a camera is used for analysing the light section in the image and recognise objects like cylinders [Grimson *et al.*, 1993] or quadrics of revolution [Tsai *et al.*, 2005]. Other alternatives consist of rotating quickly the laser plane for obtaining dense reconstructions of the object and matching it onto pre-defined models [Lin *et al.*, 1996]. The identification of objects by using these sensors is usually used in robotics as a previous step for planning manipulation operations like object grasping [Lozano-Perez *et al.*, 1987].

Simultaneous Localisation and Mapping (SLAM): structured light can be used for mobile robot self-localisation [De la Escalera *et al.*, 1996; Dubrawski and Siemiatkowska, 1998; Neira *et al.*, 1999], for map building [Kim and Cho, 2001] and for SLAM [Jung *et al.*, 2004; Surmann *et al.*, 2003]. Typically, a horizontal laser plane (static or rotating in tilt) and a camera are used for obtaining range information of the environment. In most cases these data are fused by other sensor measurements like ultrasonic sensors.

Obstacle detection/avoidance: in mobile robotics, the typical configuration for detecting obstacles consists of laser planes projecting onto the floor, or in parallel to it, and a camera as in [Weckesser *et al.*, 1995; Haverinen and Roning, 1998]. However, there are other sensors like the laser pointers and camera used in the underwater robot in [Kondo and Tamaki, 2004], the rotating laser plane and the catadioptric camera for omnidirectional perception in [Joung and Cho, 1998], or the grid projector and camera in [Le Moigne and Waxman, 1988]. Once the obstacles are detected the robot trajectory can be modified in order to avoid collisions.

Applications for industrial manipulators

In this case, we consider an industrial robot with a fixed base and several joints or moving axis and an end-effector where different types of tools can be attached. This type of robots have a great impact in the current industry.

Grasping: the operation of grasping objects or industrial pieces can benefit from the structured light as shown in [Lozano-Perez *et al.*, 1987]. In this case, laser planes linked to the robot end-effector are projected. Another example is found in [Gutsche *et al.*, 1991] where a sequence of binary patterns are projected and imaged from a deported LCD projector and camera, respectively, in order to reconstruct the object and then perform a pick and place operation with a robot arm.

Object pose estimation and robot positioning: in this case, onboard laser planes and a camera are used for determining the pose of objects. In [Albus *et al.*, 1982] a sensor formed by a camera and two laser planes was proposed. Then, in [Rutkowski *et al.*, 1987] this sensor is used for locating piecewise objects consisting of planes, spheres and cylinders. Then, a position-based visual servoing approach is used for moving the robot relative to the pieces. In [Kemmotsu and Kanade, 1995] three laser planes are used, while in [Niel *et al.*, 2004] two cameras and a grid projector are exploited.

Welding and trimming: the use of lasers planes and a camera is also used for guiding the robot tool in welding of metallic pieces [Smati *et al.*, 1987; Agapakis, 1990; Kim *et al.*, 1999]. Similarly, a light plane is used for locating, reconstructing and tracking the seam in trimming operations [Amin-Nejad *et al.*, 2003].

Machined finishing: an example of finishing operation of manufactured parts can be found in [Kwok *et al.*, 1998]. In this case, a calibrated laser plane and a camera mounted on the tool holder of a robot arm are used to collect 3D data of turbine blades. Then, the blade is automatically polished with the tool.

Surface inspection: in [Nurre *et al.*, 1988] a pattern of several parallel laser planes are projected for inspecting the quality of manufactured parts. Similarly, a grid pattern is used in [Zhang and Ma, 2000].

Medical applications

Thanks to its non-invasive nature, structured light has a great potential in medical applications. In [Nwodoh *et al.*, 1997], a camera and a laser plane are linked to the end-effector of a manipulator for obtaining 3D data from a lying human body. The aim of the application is guiding the robot end-effector for removing burnt skin tissue from the injured parts of the body. Another example is shown in [Buendía *et al.*, 1999] where a coloured grid is projected onto the back of the patient which is observed by a colour camera. The aim is to obtain a measure of the spine deviation from 3D reconstruction. Finally, structured light can also help in robotised laparascopic surgery as explained in [Krupa *et al.*, 2003].

4.4.2   Image-based visual servoing based on structured light

In the applications summarised above, structured light is typically used as an additional sensor providing 3*D* information of the robot's environment. All these applications rely on an accurate calibration of the camera and the structured light emitters in order to obtain 3*D* information through triangulation [Besl, 1988; Jarvis, 1993]. Therefore, the cited examples where 3*D* data are used for controlling the robot motion can be considered position-based approaches as in [Rutkowski *et al.*, 1987; Amin-Nejad *et al.*, 2003]. However, there are no works providing results concerning the sensitivity to image noise and camera-sensor calibration for such type of approach.

   There are few works addressing the use of the visual features provided by structured light in image-based visual servoing. Urban used two orthogonal laser planes and a camera for positioning a robot manipulator with respect to car batteries [Urban, 1990]. The visual features used were the straight lines in the image provided by the lasers. Motyl et al. [Motyl, 1992; Khadraoui *et al.*, 1996] properly investigated eye-in-hand systems with structured light emitters attached to the camera. They adapted the calculation method of the image jacobian corresponding to simple geometric primitives [Espiau *et al.*, 1992] to the case when they are provided by the projection of laser planes onto simple objects. It was shown that in this type of configurations, both the projected light and the object surface must be modelled in the image jacobian. They modelled different cases where laser planes are projected to planar and spherical objects. Their study also focuses on fixing virtual links between the camera and such objects with the aid of laser planes. Different sets of visual features extracted from the projected light were formulated like straight lines, discontinuity points and ellipse moments.

Some years later, Andreff et al. included depth control by using structured light in their 2 1/2 D approach based on lines [Andreff *et al.*, 2002]. Their control scheme was however depth-invariant. They prevented this lack of depth-control by providing a laser pointer to the camera. They first formulated the variation of the distance between the laser and its projection onto the object due to the robot's motion. The interaction matrix of this $3D$ feature was formulated taking into account a planar object. The result was equivalent to the one presented by Samson et al. [Samson *et al.*, 1991] concerning thin-field range sensors. Afterwards, they showed that the projected point lies always onto a line in the camera image, which is the unique epipolar line of the particular stereo system composed by the camera and the laser. Then, they chose as $2D$ visual feature the distance along the epipolar line between the current position of the point and a certain reference point. The variation of such feature was related to the variation of the distance between the laser and the object.

In [Ramachandram and Rajeswari, 2000], a laser with a diffracting lens emitting a 4×4 grid was linked to a robot end-effector with a camera for positioning tasks with respect to complex volumetric objects. The approach is based on bringing the robot to the desired position relative to the object and training a neural network for estimating the non-linear function relating the variation of the observed grid projected onto the object surface and different robot motions. Nevertheless, neither experimental results nor convergence conditions are provided.

More recently, Krupa et al. [Krupa *et al.*, 2003] applied visual servoing and structured light to laparoscopic surgery. In their application, two incision points are made into the patient body: one for introducing an endoscopic camera and the other to introduce a surgical instrument held by the end-effector of a robot. The camera pose with respect to the robot frame is unknown. The task consists in moving the instrument to a certain point of the observed image keeping a desired depth to the underlying organ. Due to constraints of the surgical incision and the type of instrument, only three degrees of freedom must be controlled. The instrument holder is a laser pointer which has three LEDs which are collinear with the projected spot. The depth from the surgical instrument to the organ is estimated by using the cross ratio of the projected laser dot and the three LEDs and the corresponding image points. The rotational degrees of freedom are controlled by an online identification of the image jacobian of the projected spot. On the other hand, the depth is directly controlled from the calculated estimation.

Finally, a robotic arm with eye-in-hand configuration and 5 laser planes attached to the end-effector is described in [Kahane and Rosenfeld, 2004]. The application consists in automatically adding tiles in a wall by using a suction gripper in the end-effector. A visual iterative procedure is used for guiding the robot, so that it can be considered an image-based visual servoing technique. Visual features extracted from the laser lines projected onto the wall and the already positioned tiles are used. First, the end-effector is oriented with respect to the wall so that the gripped tile gets parallel to it. Then, the lengths and orientation of the broken laser segments appearing among the gripped tile and the adjacent tiles are used for aligning and adding the new tile. However, the interaction matrix is not analytically derived and heuristics are used instead. Therefore, no analytic results about the stability of the system are available.

In this chapter a new approach to visual servoing based on structured light is proposed. Our proposal consists in placing a structured light emitter beside the robot manipulator as shown in Figure 4.2b. Contrary to the use of an onboard emitter, see Figure 4.2a, when a deported and static emitter is used, the projected marks remain static onto the object surface. The advantage of such approach is that classic image-based visual servoing can be directly applied. Furthermore we propose to project a coded pattern for providing robust visual features independently of the object appearance. Up to our knowledge, the only work using a similar setup is found in [Gutsche et al., 1991]. In this case, a video-projector is used for emitting the sequence of binary patterns proposed by Posdamer et al. [Posdamer and Altschuler, 1982]. The work by Gusche et al. relies on precisely calibrating the robot, the camera and the projector, so that the object can be reconstructed from different points of view by using the correspondences provided by the projected patterns. Then, the reconstructed points are used for planning a trajectory for the robot end-effector, which is executed in open loop. We remark that this system is a position-based approach and that visual information is not used on-line for minimising the effects of calibration errors.

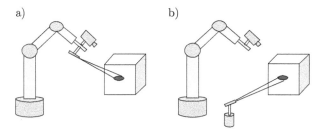

Figure 4.2: Two possible configurations of an eye-in-hand system and a structured light emitter

The remaining of the chapter is devoted to show how coded structured light can be used in a deported configuration for eye-in-hand systems. Contrary to the technique of Gutsche et al. [Gutsche et al., 1991], our proposal uses classic image-based visual servoing for reaching a pre-defined position with respect to an unknown object. A first advantage of using classic image-based visual servoing is that there is a large set of techniques for which the behaviour has been studied. Secondly, a convergence condition is available contrary to the case of the position-based approaches.

4.5 An approach based on coded structured light

Positioning a camera attached to a robot with respect to a given object by means of visual servoing is not always feasible. Despite the large variety of approaches to visual servoing that are found in the literature, there are cases where no solutions are available. For example, positioning with respect to a uniform object like the elliptic cylinder shown in Figure 4.3a becomes a hard task since no visual features can be extracted. This is an example of an object with simple appearance where visual servoing cannot be directly applied. Such situation can appear when working with industrial manipulators when positioning with respect to manufactured parts, or in mobile robots when docking with respect to walls or navigating avoiding obstacles.

On the other hand, Figure 4.3b shows a ham, a typical natural object of the agrifood industry. Note that extracting visual features from this sort of image, even if possible, requires a high computational cost. Furthermore, no a priori knowledge about the extracted features distribution onto the object surface is available. Therefore, the validity of the detected features for being tracked during the servoing is difficult to ensure.

a)   b)

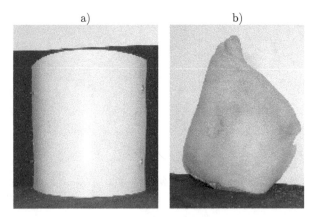

Figure 4.3: Examples of objects for which classic visual servoing is very complex.

4.5.1 Providing correspondences with a coded pattern

The study presented in Chapter 2 has shown that there are lots of coded structured light patterns providing a large number of correspondences. Among all the coded light techniques two main groups can be distinguished: the ones requiring a sequence of patterns and the ones based on a unique pattern previously referred as one-shot patterns. As pointed out by Gutsche et al. [Gutsche *et al.*, 1991], a sequence of patterns could be used in the desired state and in the initial state. Thus, a set of correspondences between the images of both robot configurations will be obtained. Then, the required motion to bring the robot from the initial configuration to the desired one could be executed in open loop. This solution, however, is not suitable because the robustness of visual servoing against calibration errors is lost since no visual feedback is included in the control law [Chaumette, 1998].

As an alternative to the use of a sequence of patterns, we propose to use a one-shot technique for providing correspondences between the desired image and the image corresponding to each iteration of the control loop. We refer to this approach as visual servoing based on deported coded structured light and the setup is represented in Figure 4.4. The emitter device is a LCD video-projector which remains separated from the robot manipulator and whose field of view is expected to illuminate the object of interest.

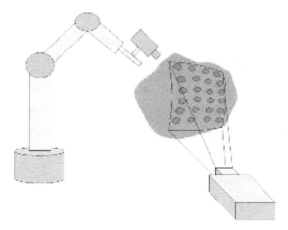

Figure 4.4: Schema of the setup for eye-in-hand visual servoing using a deported coded light projector.

The choice of the coded pattern can depend on the object, the number of correspondences that we want to get, the lighting conditions and the devices used (camera and projector). The use of LCD projectors allows the projected pattern to be changed with no cost thanks to its flexibility. The pattern chosen for this chapter could have been any one of the presented in Chapter 2 or the new one-shot pattern presented in Chapter 3. However, the choice has been taken considering a pattern for which the decoding process can be easily implemented at the rate required for visual servoing. The decoding time is not crucial in a shape acquisition setup if it can be done off-line [Zhang *et al.*, 2002]. In visual servoing, however, the time available for solving the correspondence problem between the current and the desired image must be chosen in order to not penalising the dynamics of the end-effector.

Taking into account these requirements a pattern encoded according to a m-array strategy has been chosen. Concretely, a 20×20 m-array based on the alphabet $0, 1, 2$ has been generated by using the algorithm by Morano et al. [Morano *et al.*, 1998] described in Section 2.5.3. The resulting m-array has window property 3×3 so that any sub-array of this size appears as maximum once in the whole array. The resulting array is shown in Figure 4.5. A pattern of dimensions 768×768 pixels has been designed containing 20×20 dots coloured according to the m-array. Symbol 0 has been mapped to blue, 1 to green and 2 to red. The resulting pattern is presented in Figure 4.6.

4.5.2   Robust segmentation and decoding of the pattern

The aim of projecting the encoded pattern onto the object of interest is to easily find matchings between images taken from different points of view. The coding scheme included in the pattern allows a list of points to be identified and labelled in each image. Then, the matching between images can be directly done by using the labels of the decoded points in each image. The process of identifying the projected pattern onto the image correspond to the projected pattern is referred as pattern segmentation. Once the elements of the pattern are segmented, they can be decoded taking into account the coding strategy used to generate the pattern. Given the m-array pattern consisting of coloured dots in Figure 4.6, the segmentation and decoding procedures are detailed in the following paragraphs.

The pattern segmentation is done according to the following steps:

- **RGB binarisation:** the pattern projects high intensity red, green and blue spots on a black background. The iris of the camera's objective can be closed enough for only perceiving the coloured spots projected onto the object of interest. Then, the colour of every pixel of the image can be modified according to the following function

$$
pixel(x, y) = \begin{cases} (1, 0, 0) & if \ max(R, G, B) = R \ and \ R > threshold \\ (0, 1, 0) & if \ max(R, G, B) = G \ and \ G > threshold \\ (0, 0, 1) & if \ max(R, G, B) = B \ and \ B > threshold \end{cases} \quad (4.14)
$$

  where R, G and B are the original red, green and blue level of $pixel(x, y)$ and threshold is the minimum intensity level for being considered part of the projected pattern. With this per-channel binarisation, the imaged spots are segmented to the most likely colour and regions of the image not highly illuminated, i.e. not illuminated by the pattern, are removed.

- **Spot segmentation:** in this step the gravity centre of every coloured spot are located. First, the RGB image provided by the previous step is binarised according

0	2	0	1	1	1	1	1	2	1	1	0	1	0	1	1	0	1	1	2
2	1	2	0	0	0	0	0	0	0	0	1	0	1	0	2	1	0	0	1
0	0	1	1	1	2	2	2	2	2	1	0	2	0	1	1	0	2	1	0
1	2	2	2	2	0	0	0	1	1	0	2	1	2	2	2	2	0	0	1
2	1	1	0	0	2	1	2	2	0	1	0	2	0	0	1	0	1	2	0
0	2	0	2	1	0	0	1	0	1	2	1	1	2	1	0	2	2	0	2
2	1	1	0	0	1	1	2	2	2	0	2	0	0	0	2	0	0	1	1
0	2	2	1	2	2	0	0	1	1	1	1	1	2	2	1	2	1	2	2
1	0	0	0	0	0	1	1	2	2	2	2	0	0	0	0	0	2	0	0
0	1	2	1	1	1	2	0	1	0	1	1	2	2	1	1	2	0	2	2
1	0	0	2	2	2	0	1	0	1	2	0	0	0	2	0	0	1	1	0
0	1	1	0	1	0	2	2	1	0	0	2	1	1	0	2	2	2	0	2
1	2	0	1	2	1	1	0	2	1	1	1	0	2	1	0	0	1	1	1
2	0	2	2	1	2	2	1	1	2	2	0	2	0	2	1	1	2	2	0
0	1	0	0	2	1	1	2	0	0	0	1	0	1	1	0	0	0	1	1
1	0	1	1	1	0	0	1	1	1	2	0	2	0	2	1	1	1	0	2
0	1	2	2	2	1	2	2	0	2	1	2	0	2	1	2	0	2	2	0
2	0	0	0	1	2	1	0	2	0	2	0	1	0	0	1	2	0	0	1
0	1	2	1	2	0	2	2	1	2	0	1	2	1	1	0	0	1	1	0
1	2	1	2	1	1	0	1	2	1	1	2	1	0	2	1	2	2	0	2

Figure 4.5: 20×20 m-array based on 3 elements and 3×3 window property.

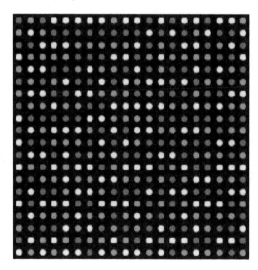

Figure 4.6: Pattern generated according the algorithm by Morano et al. and the 20×20 m-array.

to its luminance channel. Then, a region growing algorithm is used for locating every blob. Some blobs are rejected taking into account their compactness and their area.

According to Chapter 2, once the pattern spots are located in the image, the following decoding process starts:

- **Adjacency graph:** for every spot, the four closest spots in the four main directions are searched. With this step the 4-neighbourhood of each spot is located. These neighbours are used to complete the 8-neighbourhood of every spot.

- **Graph consistency:** in this step, the consistency of every 8-neighbourhood is tested. For example, given a spot, its north-west neighbour must be the west neighbour of its north neighbour, and at the same time, the north neighbour of its west neighbour. These consistency rules can be extrapolated to the rest of neighbours corresponding to the corners of the 8-neighbourhood. Those spots not respecting the consistency rules are removed from the 8-neighbourhood being considered.

- **Spot decoding:** the process of decoding consists in, for every spot having a complete 8-neighbourhood, its colour and the colours of its neighbours are used for identifying the spot in the original pattern. In order to speed up this search, a look up table is recommended.

In Figure 4.7 three results of the decoding process are presented. The original images taken by the camera with the overprinted labels of the decoded points are shown. In all the cases, the processing time required for locating and decoding the spots was about 30 ms, which is lower than the acquisition period with a CCIR format camera (40 ms).

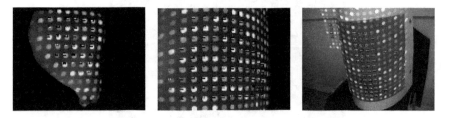

Figure 4.7: Segmentation and decoding of the pattern. a) Projection onto a ham. b) Projection onto a white elliptic cylinder. c) Projection onto a textured elliptic cylinder.

4.5.3 Task definition and control loop

The projection of the m-array pattern allows a set of correspondences to be found between the image perceived by the eye-in-hand system and the pattern. Decoded points in images taken from different points of view can be easily and robustly matched thanks to the coding strategy included in the pattern. This allows us to easily define a robot positioning task with respect to objects being illuminated by the LCD projector.

The approach consists in moving the robot to the desired pose and take an image with the camera attached to its end-effector. From this desired image, the algorithm decoding the pattern is able to recover a list of points which are corresponded to the points of the projected pattern. Then, for any other robot pose, if the object appears in the image, the same process can be done in order to obtain another list of decoded points. Once a set of points are matched between the initial image and the desired image, the robot can be guided through visual servoing in order to reach the desired configuration. The schema of this control strategy is represented in Figure 4.8.

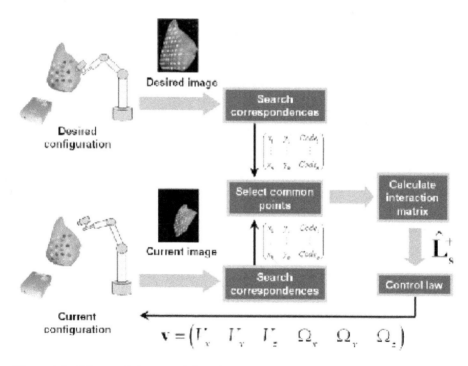

Figure 4.8: Schema of the control loop based on common points detected in both the current and the desired image.

Visual servoing requires to define a set of visual features extracted from the images. In this case, we have chosen to build a feature vector with the normalised coordinates of the common points appearing both in the current and desired image. Therefore, $\mathbf{s}$ is a $k \times 1$ vector of the form

$$\mathbf{s} = (x_1, y_1, x_2, y_2, ..., x_k, y_k) \qquad (4.15)$$

Note that there are other alternatives like the image moments of the points distribution [Tahri and Chaumette, 2004].

Let us now define the task function and control law taking into account this set of visual features. In most cases, more than 6 common points will be detected in the desired and current image due to the large amount of projected spots ($20 \times 20 = 400$). Otherwise, the task function and control law can be easily adapted following the guidelines recalled in Section 4.2. For the case of k points matched between the current and the desired image, if $k > n$, being $n = 6$ the number of degrees of freedom controlled, a possible choice for $\mathbf{W}$ is

$$\mathbf{W} = \mathbf{I_6} \tag{4.16}$$

and then, the combination matrix in (4.4) is

$$\mathbf{C} = \widehat{\mathbf{L_s}}^+ \tag{4.17}$$

being $\widehat{\mathbf{L_s}}^+$ the pseudoinverse of a model or approximation of the interaction matrix of $\mathbf{s}$. Therefore, the task function is

$$\mathbf{e} = \widehat{\mathbf{L_s}}^+ (\mathbf{s} - \mathbf{s}^*) \tag{4.18}$$

Since $\mathbf{s}$ is based on image points, it is built by stacking, as many times as matched points are available, the interaction matrix of a normalised point which is [Feddema *et al.*, 1991; Espiau *et al.*, 1992]

$$\mathbf{L_x} = \begin{pmatrix} -1/Z & 0 & x/Z & xy & -(1+x^2) & y \\ 0 & -1/Z & y/Z & 1+y^2 & -xy & -x \end{pmatrix} \tag{4.19}$$

Therefore, $\mathbf{L_s}$ has the form

$$\mathbf{L_s} = \begin{pmatrix} -1/Z_1 & 0 & x_1/Z_1 & x_1 y_1 & -(1+x_1^2) & y_1 \\ 0 & -1/Z_1 & y_1/Z_1 & 1+y_1^2 & -x_1 y_1 & -x_1 \\ \vdots & \vdots & \vdots & \vdots & \vdots & \vdots \\ -1/Z_k & 0 & x_k/Z_k & x_k y_k & -(1+x_k^2) & y_k \\ 0 & -1/Z_k & y_k/Z_k & 1+y_k^2 & -x_k y_k & -x_k \end{pmatrix} \tag{4.20}$$

Note that the real interaction matrix depends on the depth distribution of the points which is generally unknown. The model of interaction matrix $\widehat{\mathbf{L_s}}$ used in the control law is defined as the matrix evaluated at the desired state $\mathbf{L_s^*}$ by using the normalised coordinates from the desired image (x^*, y^*) and assuming that in the desired configuration the points are coplanar at a depth Z^*. This model of matrix is hereafter noted as $\widehat{\mathbf{L_s^*}}$. This type of approximation is usually sufficient as already seen in Section 4.2.2.

The proportional control in (4.11) is then

$$\mathbf{v} = -\lambda \left(\widehat{\mathbf{L_s^*}}^+ \widehat{\mathbf{L_s^*}} \right)^+ \widehat{\mathbf{L_s^*}}^+ (\mathbf{s} - \mathbf{s}^*) \tag{4.21}$$

Assuming that $\widehat{\mathbf{L}_s^*}$ is a $k \times 6$ matrix of rank 6 then

$$\widehat{\mathbf{L}_s^*}^+ \widehat{\mathbf{L}_s^*} = \mathbf{I}_6 \tag{4.22}$$

so that the control law can be modelled as

$$\mathbf{v} = -\lambda \widehat{\mathbf{L}_s^*}^+ (\mathbf{s} - \mathbf{s}^*) \tag{4.23}$$

Note that if $\widehat{\mathbf{L}_s^*}$ is near $\mathbf{L_s}$ the convergence is ensured thanks to the positivity condition in (4.13).

4.6 Experimental results

This section shows two experiments using the approach presented in this chapter. The experiments have been done with a six-degrees-of-freedom robot manipulator formed by three translational axis and three rotational joints. A colour camera has been attached to the end-effector of the robot. The camera sensor has squared pixels of $8.3\mu m \times 8.3\mu m$ and a focal length of 8.5 mm. The images are digitised by a Corona-II card at 768×582 pixels.

For the experiments, a LCD projector has been positioned about 1 m aside the robot. The focal has been set so that the pattern gets acceptably focused in a range of distances between 1.6 and 1.8 m in front of the projector. In this range of distances the objects are placed for realising the experiments.

Two experiments are now presented. In both cases non-textured objects for which classic visual servoing is unable to work have been used. The first is a planar object while the second consists of an elliptic cylinder in order to test the approach in front of objects exhibiting a curved surface.

4.6.1 First experiment: planar object

Firstly, a large non-textured white plane has been positioned in front of the robotic cell. The desired position of the robot has been defined so that the camera on the end-effector gets parallel to the plane at a distance of $Z^* = 90$ cm.

In order to learn the desired position a flat target with 4 landmarks forming an square of known dimensions has been attached to the plane as shown in Figure 4.9. Then, classic $2D$ visual servoing has been used for getting the camera at the desired distance and parallel to the plane. Once the desired position has been reached and stored, the target has been removed and the projector has been turned on so that the coded pattern is projected onto the object plane as shown in Figure 4.10.

The desired image corresponding to this robot configuration is shown in Figure 4.11a. In this image a total number of 370 coloured spots out of 400 have been successfully

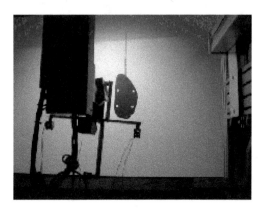

Figure 4.9: Target used for positioning the robot parallel to a plane.

decoded. Afterwards, the robot end-effector has been displaced −5 cm along its X axis, 10 cm along Y, −20 cm along Z, and rotations of −15° about X and −10° about Y have been applied. The image perceived in this configuration is shown in Figure 4.11b. In this case, the number of decoded points is 361. Matching points between the initial and the desired images is straightforward thanks to the decoding process of the pattern. Both images shown in Figure 4.11 are the ones resulting of the RGB binarisation procedure described in Section 4.5.2.

The goal is then to move the camera back to the desired position by using visual servoing. At each iteration, the visual features set **s** in (4.15) is filled with the matched points between the current and the desired image. The model of interaction matrix used both in the task function definition (4.18) and the control law (4.23) are computed at each iteration with the desired image point coordinates corresponding to the coloured dots appearing both in the current and desired image. Furthermore, the depth of all the points in the desired configuration has been set to its right value of $Z^* = 90$ cm. The result of the servoing is presented in Figure 4.11c-d. Concretely, the camera velocities generated by the control law are plotted in Figure 4.11c. Note that the norm of the task function decreases at each iteration as shown in Figure 4.11d. As can be seen, the behaviour of both the task function and the camera velocities is satisfactory and the robot reaches the desired position with no problems.

4.6.2   Second experiment: elliptic cylinder

In the second experiment a non-planar object has been used. Concretely, the elliptic cylinder shown in Figure 4.12 has been positioned in the workspace. In this case, the desired position has been chosen so that the camera points to the object's zone of maximum curvature with a certain angle and the distance between both is about 60 cm. The desired image perceived in this configuration is shown in Figure 4.13a. The number of successfully decoded points is 160. Then, the robot end-effector has been displaced −20 cm along X,

Figure 4.10: First experiment: projection of the coded pattern onto a planar object

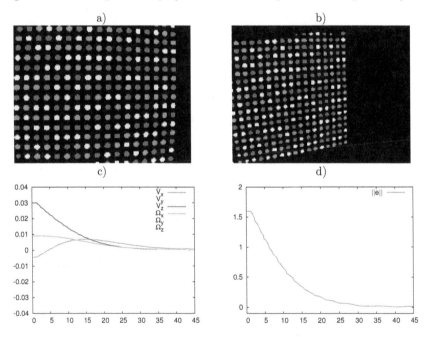

Figure 4.11: First experiment: planar object. a) Desired image. b) Initial Image. c) Camera velocities (ms/s and rad/s) vs. time (in s). d) Norm of the task function vs. time (in s).

−20 cm along Y and −30 cm along Z. Afterwards, rotations of −10° about X, 15° about Y and 5° about Z have been effectuated. This defines the initial position of the robot end-effector. The image perceived in this configuration is shown in 4.13b. In this case, the number of decoded points is 148.

The results of the visual servoing are plotted in Figure 4.13c-d. The desired image is reached again at the end of the servoing. However, the camera velocities generated by the control law are more noisy and less monotonic than in the previous experiment. Furthermore, the task function takes more time to cancel. We suppose that this is due to the fact that the modelling assumptions taken in this experiment were quite far from the real case. Note that the model of interaction matrix used in the control law is not only based on the point distribution of the desired image, but also considers that all the points are coplanar at depth $Z^* = 60$ cm. Since the object has a high curvature, the points used in the task function are non-coplanar so that the interaction matrix used is false. Recently, Malis et Rives proved that the depth distribution of a cloud of points used in visual servoing plays an important role in the stability of the system [Malis and Rives, 2003]. Nevertheless, this example confirms that visual servoing is usually quite robust against modelling errors [Chaumette, 1998]. Furthermore, during the robot motion some of the pattern points were occluded by the robot itself. Therefore, the control law has shown to be robust against occlusions.

4.7 Conclusions

This chapter proposes to use deported structured light in order to provide visual features and to perform classic image-based visual servoing for robot positioning. The motivation of our approach comes from the fact that most part of applications in robotics using structured light are based on $3D$ data provided by the sensor. There are really few works using image-based visual servoing based on features provided by structured light. The use of classic image-based visual servoing has the advantage that there are a lot of techniques available and their behaviour has been pretty studied. Furthermore, when using image-visual servoing a convergence condition is available, something which does not happen when using $3D$ visual servoing.

Our approach is based on attach a camera to the end-effector of the robot for perceiving its environment. Furthermore, a LCD projector is placed aside the robot for illuminating the working area with a coded pattern. One of the aims of projecting light patterns is obtaining visual features when dealing with uniform or non-textured objects. Furthermore, including a coding strategy in the pattern allows correspondences between the image and the pattern to be robustly recovered as ambiguities are removed. Thus, matching visual features between the initial image, intermediate images captured during the robot motion, and the desired image becomes straightforward.

A pattern containing a m-array of coloured spots has been used for illustrating this approach. The choice of this pattern has been made taking into account its easy segmentation and fast decoding, which fits on the visual servoing requirements of sampling rate. Concretely, the computing time has kept in about 35 ms per iteration. To our knowledge,

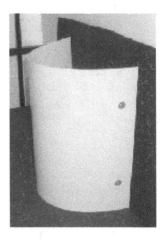

Figure 4.12: Second experiment: projection of the coded pattern onto an elliptic cylinder.

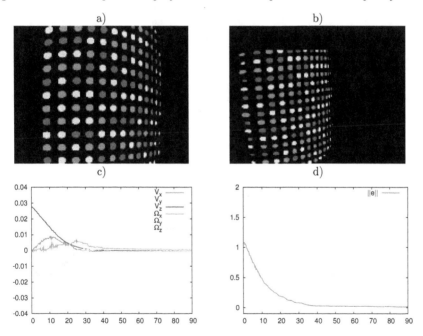

Figure 4.13: Second experiment: elliptic cylinder. a) Desired image. b) Initial Image. c) Camera velocities (ms/s and rad/s) vs. time (in s). d) Norm of the task function vs. time (in s).

this is the first work using coded structured light in a visual servoing framework. Therefore, we consider this approach a first step which shows the potentiality of coded light in
visual servoing applications.

A classic image-based approach based on points provided by the coded pattern has
been used. Experiments have shown that good results are obtained when positioning
the robot with respect to planar objects. Furthermore, thanks to the large number of
correspondences provided by the coded pattern, the system has shown to be robust even
in presence of eventual occlusions. On the other hand, the results when using non-planar
objects are not so good. This is a well known problem in classic $2D$ visual servoing. For
example, assumptions about the depth distribution of the points play an important role
in the convergence [Malis and Rives, 2003]. The advantage of using coded light is that the
pattern can be changed in order to obtain other visual features so that any of the existing
$2D$ visual servoing approaches can be used. Indeed, the use of different types of visual
features will change the behaviour of the robot [Chaumette, 1998]. For example, with the
current pattern, image moments of the point distribution [Tahri and Chaumette, 2004] or
the extended-$2D$ visual servoing [Schramm et al., 2004] could be used in order to improve
the behaviour. For improving the behaviour of the system when positioning with respect
to non-planar objects, the depths could be estimated by triangulation, even if it would
require to calibrate the projector. Another solution would be to use $2D$ 1/2 visual servoing
as in [Malis and Chaumette, 2000].

The main constraint of the current approach is that the pattern used is not rotation
invariant. This means that in order to properly decode the pattern it cannot appear
too much rotated in the image. Hence, if the camera is considerably rotated around the
optical axis from the initial to the desired position, no correspondences can be found. This
problem does not appear in the presented experiments since no large rotations around the
optical axis were used when defining both the desired and the initial robot configuration.
In order to remove this constraint a pattern which is rotation invariant can be used, as
the one proposed in [Salvi et al., 1998] at the expense of a more costly segmentation and
decoding procedure.

The work developed in this chapter can be improved with new research perspectives
proposed in Chapter 6.

Finally, we insist on the fact that structured light allows us to choose the visual features
which will be used in the control law. In the following chapter we show that a good choice
of the projected pattern can optimise the control law of the visual servoing approach.

Chapter 5

A structured light sensor for plane-to-plane positioning

In this chapter we address a robot positioning task by using an onboard structured light emitter. The case of a deported structured light emitter has been studied in the previous chapter. The onboard case is pretty challenging as the modelling of the variation on the visual features due to the camera motion is more complex. Therefore, an important modelling effort is developed in this chapter. A dedicated structured light sensor for plane-to-plane positioning is proposed. The sensor has been designed in order to optimise the visual control loop and to obtain nice properties like decoupling, stability and good camera trajectory. Several sets of visual features are presented providing stability analysis, simulations and experimental results.

5.1 Introduction

The previous chapter has shown that a deported source of coded structured light can provide robust visual features to complex or non-textured objects. This can be a very useful solution in robotic cells for manipulating static objects for which extracting its own visual features is difficult.

On the other hand, in case of certain robotic tasks, moving objects or in mobile robotics, the use of a deported source of light can be unappropriated. In these cases, the use of an onboard structured light emitter linked to the camera seems a better solution. However, the current dimensions and weight of LCD projectors prevent them to be placed on the end-effector of robot manipulators or in some mobile robots. This technologic constraint is expected to be removed in a near future by miniaturising the devices and diminishing their cost. By the time, as already mentioned in Section 4.4.1, most part of robotics applications using onboard structured light are based on laser technology. In this case, the available patterns are dots, lines, circles and grids. However, most part of applications combining structured light and cameras in robotics use these devices for obtaining range measures by triangulation. For example, the underwater robot in [Kondo and Tamaki, 2004] uses two laser pointers and a camera for obstacle avoidance and docking maneuvering. In [Sun *et al.*, 2004], a climbing robot uses two laser pointers and a camera for obtaining its orientation with respect to the window pane that it is cleaning. Finally, we can mention the robot manipulators in [Clocksin *et al.*, 1985; Amin-Nejad *et al.*, 2003] taking profit of onboard laser planes and cameras for reconstructing surfaces for welding and trimming operations. It should be notice that these examples and a lot of applications taking profit of structured light rely on an accurate calibration of the devices for obtaining 3D reconstructions. At the same time, we note that, in most cases, the task to perform could be formulated as a visual servoing task. Then, the calibration of the devices could be softened thanks to the closed-loop control based on visual feedback. As a clear example, we can cite the robot in [Kahane and Rosenfeld, 2004] for automatic tile mosaicking. In this application the robot adds tiles in a wall by visually guiding the gripper by using an onboard camera and five laser planes. Even if a control loop based on visual feedback is used, the analytic expression of the interaction matrix related to the visual features is not calculated. We think that with a stricter formulation of the visual task the system behaviour could be easily improved.

Although the important potentiality on performing positioning or navigating tasks in robotics by combining onboard structured light and visual servoing, there exist few works exploiting this type of approach. For example, Andreff et al. [Andreff *et al.*, 2002] used an onboard laser pointer in their eye-in-hand configuration for depth control. Similarly, Krupa et al. [Krupa *et al.*, 2003] coupled a laser pointer to a surgical instrument in order to control its depth to the organ surface, while both the organ and the laser are viewed from a static camera. Urban et al. [Urban *et al.*, 1994] used two onboard laser planes an a camera for positioning a 2 degrees of freedom (dof) robot end-effector with respect to car batteries. In general, most of the applications only take profit of the emitted light in order to control one or two dofs and to make easier the image processing. There are few works addressing the issue of controlling several dofs by using visual features extracted from the projected structured light. The main contribution in this field is due to Motyl et al. [Motyl, 1992; Khadraoui *et al.*, 1996], who modelled the interaction matrix of the visual features obtained when projecting laser planes onto planar objects and spheres.

The objective of this chapter is to implement a positioning task with visual servoing and using onboard structured light in an eye-in-hand configuration. In this case, and contrary to the previous chapter, the structured light will be used not only for providing visual features in non-textured or uniform objects, or under adverse lighting conditions, but also for showing that an adequate design of the structured light sensor can be used for obtaining a robust control law. By robust we mean that the system is able to reach the goal from any initial state even if modelling and calibration errors have been made. The robustness can be checked through stability analysis. There are several design strategies which are intended to produce robust control laws like designing a decoupled control law and producing smooth camera velocities.

Certainly, during the last years, many works in visual servoing have been directed on developing approaches for which the convergence of the system is largely ensured even if the initial position is far from the desired one [Andreff *et al.*, 2002; Malis and Chaumette, 2002; Schramm *et al.*, 2004]. As mentioned before, a suitable design strategy consists in searching for decoupled visual features, so that each one only controls one dof. Even if such control design seems to be out of reach, there are several works concerning the problem of partially decoupling dofs. For example, hybrid techniques are based on controlling rotational dof in the cartesian space while the translational ones are controlled by image information [Deguchi, 1997; Malis and Chaumette, 2002]. However, they require partial pose estimation of the object at each iteration. On the other hand, some pure image-based techniques have succeed to decouple rotational dof from translational ones near the desired state [Corke and Hutchinson, 2001; Tahri and Chaumette, 2004]. Concerning the stability analysis, most part of techniques for which it has been possible to find analytical conditions are hybrid approaches like in [Andreff *et al.*, 2002; Malis and Chaumette, 2002] or more recently, the extended-2D visual servoing [Schramm *et al.*, 2004]. Usually, the global stability analysis of pure image-based techniques is too complex even in absence of calibration errors.

Another important research topic in image-based visual servoing is to improve the camera trajectory in the cartesian space. It is well known that even if an exponential decrease of the visual error is achieved, it does not necessarily imply a suitable camera trajectory. This is mainly due to strong non-linearities in the image jacobian. Important efforts have been done in order to improve the mapping from the feature space to the camera velocities [Mahony *et al.*, 2002; Tahri and Chaumette, 2004].

In this chapter we contribute to these research topics by using a dedicated structured light emitter which allows the plane-to-plane positioning task to be optimised. This task consists in moving the camera so that it gets parallel to a planar object. This classic task has been chosen in order to show that structured light can be useful for obtaining nice properties like decoupling, stability and good camera trajectory.

First of all, we develop a method for a better estimation of the object plane para-meters by using the structured light pattern. Then, several image-based approaches are formulated leading to a decoupled approach which is less time consuming and less sensi-tive to image noise than a position-based approach based on the object pose estimation. Furthermore, its robustness against calibration errors is demonstrated analytically and experimentally. In addition to this, a linear map from the task function to the camera velocities is made, producing a suitable camera trajectory.

The chapter is structured as follows. The structured light sensor proposed to fulfill the plane-to-plane virtual link and its modelling is presented in Section 5.2. Then, Sec-tion 4.2 reviews the plane-to-plane task function definition, presents the control law and the procedure used to analyse its stability. Afterwards, four visual servoing approaches exploiting the projected light are presented. First, Section 5.4 deals with several position-based approaches based on reconstructing the object pose by triangulation. Then, a simple 2D approach based on image points coordinates is shown in Section 5.5. After that, in Section 5.6 a 2D approach based on the area corresponding to the projected pattern and combinations of angles extracted from the image is analysed. The last approach is based on a robust non-linear combination of image points which is presented in Section 5.7. Some experiments showing the behaviour obtained with each one of the proposed approaches are shown in Section 5.8. Finally, despite the sensor has been developed to cope with planar objects, the behaviour of the last approach in presence of several non-planar objects is tested in Section 5.10. The chapter ends with conclusions.

5.2   A proposal of structured light sensor for plane-to-plane positioning

We present an eye-in-hand system with a structured light sensor attached to the camera. The goal of the task here addressed consists in positioning the camera parallel to a planar object. Such type of task, namely plane-to-plane positioning, aims to fix a virtual link between the camera image plane and the object plane. With this classic task we aim to demonstrate that using a suitable structured light emitter, the performance of the visual servoing scheme can be optimised in terms of decoupling, stability and camera trajectory.

The structured light sensor is based on laser pointers since they are low-cost and easily available. Theoretically, three non-collinear points are enough to recover the parameters of the equation of a planar object. Consequently, we initially designed a sensor composed of three laser pointers. Nevertheless, we found that better results can be obtained by using four laser pointers in order to decouple visual features.

Therefore, the structured light sensor that we propose consists of four laser pointers attached to a cross-shaped structure as shown in Figure 5.1a. The direction of the lasers have been chosen to be equal so that four points are orthogonally projected to the planar object, see Figure 5.1b. This causes that the projected shape enclosed by the four points is invariant to the distance between the object and the laser-cross. This invariance will be very useful as will be shown in following sections. The symmetric distribution of the lasers in the cross structure will be also useful for decoupling the visual features.

Consequently, the model of the proposed sensor is as follows. The laser-cross has its own frame {L} so that all the lasers have the same orientation vector $^{L}\mathbf{u} = (0, 0, 1)$. Furthermore, the lasers are placed symmetrically so that two of them lie on the X_L axis and the other two on the Y_L axis. All the lasers are positioned at a distance L from the origin of the laser-cross frame. The structured light sensor is modelled assuming that it is ideally attached to the camera of the robot manipulator as shown in Figure 5.1b . As can be seen, in this model, the cross-laser frame perfectly coincides with the camera frame, so that the structured light sensor is placed just in the camera origin and the lasers point toward the direction of the camera optical axis. Whenever the camera and the planar object are parallel, the projected laser points exhibit a symmetric distribution onto the object surface and also in the camera image, see Figure 5.2, which will allow us to find decoupled visual features.

Note that these assumptions have been only taken for modelling issues. However, it is perhaps not always possible to perfectly align the laser-cross with the camera frame because of the structure of the robot or because the optical centre position is not exactly known. That is why the study of the robustness against misalignments between the camera and the laser-cross will be a key point when analysing the different approaches presented in this chapter.

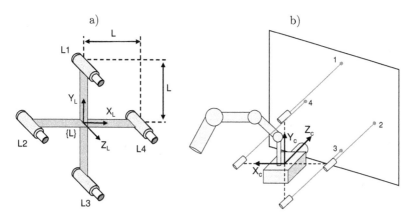

Figure 5.1: System architecture. a) The proposed structured light sensor. b) Ideal configuration of the robot manipulator, camera and structured light sensor.

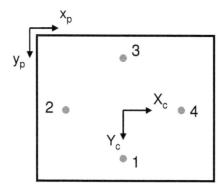

Figure 5.2: Camera image when it is parallel to the object at a given depth.

Next section presents the modelling of a laser pointer and the image jacobian corresponding to the projected point onto a planar object. Afterwards, the whole model of the proposed structured light emitter is presented under ideal conditions and different types of calibration errors.

5.2.1   Laser pointer modelling

In visual servoing, given a set of visual features **s** extracted from an image, its variation due to the relative camera-object velocity (*kinematic screw*) is expressed in the well known equation

$$\dot{\mathbf{s}} = \mathbf{L_s}\mathbf{v} \tag{5.1}$$

being $\mathbf{v} = (V_x, V_y, V_z, \Omega_x, \Omega_y, \Omega_z)$ the camera velocity screw assuming a static object, and $\mathbf{L_s}$ the image jacobian known as *interaction matrix*.

Given a 3D point $\mathbf{X} = (X, Y, Z)$ fixed to the observed object, its normalised coordinates $\mathbf{x} = (x, y)$ resulting of the perspective projection $\mathbf{x} = \mathbf{X}/Z$ are the most widely used features in image-based visual servoing. The interaction matrix of a fixed point of coordinates (x, y) is [Espiau *et al.*, 1992; Feddema *et al.*, 1991]

$$\mathbf{L_x} = \begin{pmatrix} -1/Z & 0 & x/Z & xy & -(1+x^2) & y \\ 0 & -1/Z & y/Z & 1+y^2 & -xy & -x \end{pmatrix} \tag{5.2}$$

note that the only 3D information included in the interaction matrix is the depth of the point which appears in the translational components.

The analog case when working with structured light consists in using a laser pointer so that the intersection of the laser with the object produces also a point $\mathbf{X}$ as shown in Figure 5.3. When the laser pointer is linked to the camera in an eye-in-hand configuration, the time variation of the observed point $\mathbf{x}$ depends also on the geometry of the object since $\mathbf{X}$ is not a static physical point. Therefore, some modelling of the object surface must be included in the interaction matrix.

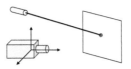

Figure 5.3: Case of a laser pointer and a planar object.

In this work we focus on the case of planar objects, which are modelled according to the following explicit equation

$$\underline{\mathbf{n}}^{\top}\mathbf{X} + D = 0 \tag{5.3}$$

being $\underline{\mathbf{n}} = (A, B, C)$ the unitary normal vector to the plane and D its distance to the origin of the camera frame. Hereafter, we take the convention that $C > 0$ which implies that $D < 0$ since the object is in front of the camera.

Motyl et al. were the first on formulating the interaction matrix of a projected point onto a planar object [Khadraoui *et al.*, 1996]. In their case, the projected point was the result of the intersection of two laser planes with the planar object. Note that a laser pointer (straight line in the space) can be modelled as the intersection of two planes, so that it is equivalent to projecting two intersecting laser planes. The interaction matrix proposed by Motyl et al. has the disadvantage of depending on 12 $3D$ parameters: 4 parameters for every one of the two laser planes (normal vector and distance to the origin) plus 4 parameters for the planar object. Furthermore, the explicit depth Z of the point does not appear in the interaction matrix.

In a more natural way, the laser pointer can be modelled with a vectorial equation as follows (all the parameters are expressed in the camera frame)

$$\mathbf{X} = \mathbf{X}_r + \mu \underline{\mathbf{u}} \tag{5.4}$$

where $\underline{\mathbf{u}} = (u_x, u_y, u_z)$ is an unitary vector defining the laser direction, $\mathbf{X}_r = (X_r, Y_r, Z_r)$ is any reference point belonging to the straight line, and μ is the distance from $\mathbf{X}_r$ to $\mathbf{X}$.

By deriving the above expression and taking into account that both $\mathbf{X}_r$ and $\underline{\mathbf{u}}$ do not change when the camera moves we find that the time derivative of the projected point is

$$\dot{\mathbf{X}} = \dot{\mu} \underline{\mathbf{u}} \tag{5.5}$$

then, deriving the normalised coordinates $\mathbf{x}$ and using the above result we can write

$$\dot{\mathbf{x}} = \frac{\dot{\mathbf{X}}}{Z} - \frac{\mathbf{X}}{Z^2}\dot{Z} = \frac{1}{Z}\dot{\mu}\underline{\mathbf{u}} - \frac{\mathbf{x}}{Z}\dot{Z} \tag{5.6}$$

from (5.5) we have that $\dot{Z} = \dot{\mu}u_z$ so that

$$\dot{\mathbf{x}} = \frac{\dot{\mu}}{Z}(\underline{\mathbf{u}} - \mathbf{x}u_z) \tag{5.7}$$

In order to calculate $\dot{\mu}$ let first express μ in function of the $3D$ parameters. By substituting $\mathbf{X}$ in (5.3) by its expression in (5.4) we have

$$\mu = -\frac{1}{\underline{\mathbf{n}}^\top \underline{\mathbf{u}}}(\mathbf{n}^\top \mathbf{X}_r + D) \tag{5.8}$$

deriving this expression we obtain

$$\dot{\mu} = -\frac{1}{\underline{\mathbf{n}}^\top \underline{\mathbf{u}}}(\dot{\mathbf{n}}^\top \mathbf{X}_r + \dot{D}) + \frac{\mathbf{n}^\top \mathbf{X}_r + D}{(\underline{\mathbf{n}}^\top \underline{\mathbf{u}})^2}\dot{\mathbf{n}}^\top \underline{\mathbf{u}} \tag{5.9}$$

which can be reduced to

$$\dot{\mu} = -\frac{1}{\underline{\mathbf{n}}^\top \underline{\mathbf{u}}}(\dot{\mathbf{n}}^\top (\mathbf{X}_r + \mu\underline{\mathbf{u}}) + \dot{D}) \tag{5.10}$$

and finally, applying (5.4) the time derivative of μ is

$$\dot{\mu} = -\frac{1}{\underline{\mathbf{n}}^\top \underline{\mathbf{u}}}(\dot{\mathbf{n}}^\top \mathbf{X} + \dot{D}) \tag{5.11}$$

Taking into account the time derivatives of the planar object parameters [Urban *et al.*, 1994]

$$\begin{pmatrix} \dot{\underline{n}} \\ \dot{D} \end{pmatrix} = \begin{pmatrix} \mathbf{0}_{3\times3} & [\underline{n}]_\times \\ \underline{n}^\top & \mathbf{0}_{1\times3} \end{pmatrix} \mathbf{v} \tag{5.12}$$

where $[\underline{n}]_\times$ is the antisymmetric matrix associated to vector $\underline{n}$, the interaction matrix of μ is

$$\mathbf{L}_\mu = -\frac{1}{\underline{n}^\top \underline{u}} \left(\mathbf{n}^\top \ (\mathbf{X} \times \underline{n})^\top \right) \tag{5.13}$$

The equivalence of this formula to the one presented by Samson et al. [Samson *et al.*, 1991] and the one provided by Andreff et al. [Andreff *et al.*, 2002] is shown in appendix A.

By using the time derivative of μ, Equation (5.7) can be rewritten as follows

$$\dot{\mathbf{x}} = -\frac{1}{\underline{n}^\top \underline{u}} (\underline{u} - x u_z) \frac{(\dot{\underline{n}}^\top \mathbf{X} + \dot{D})}{Z} \tag{5.14}$$

note that the time derivative of $\mathbf{x}$ is expressed in function of 7 3D parameters, namely Z, $\underline{n}$ and $\underline{u}$. The only parameters concerning the laser configuration are the components of its direction vector $\underline{u}$. This result can be still improved by expressing $\underline{u}$ as follows

$$\underline{u} = (\mathbf{X} - \mathbf{X}_r)/\|\mathbf{X} - \mathbf{X}_r\| \tag{5.15}$$

applying this expression in (5.14) and after some developments $\dot{\mathbf{x}}$ becomes

$$\dot{\mathbf{x}} = \frac{(\mathbf{X}_r - \mathbf{x} Z_r)}{\underline{n}^\top (\mathbf{X} - \mathbf{X}_r)} \left(\frac{\dot{D}}{Z} + \dot{\underline{n}}^\top \mathbf{x} \right) \tag{5.16}$$

Note that the expression does not longer depend on the orientation of the laser $\underline{u}$ but on its reference point $\mathbf{X}_r$. Furthermore, if the reference point $\mathbf{X}_r$ is chosen as $\mathbf{X}_r = \mathbf{X}_0 = (X_0, Y_0, 0)$, which corresponds to the intersection of the straight line modelling the laser and the plane $Z = 0$ of the camera frame, the expression simplifies to

$$\dot{\mathbf{x}} = \frac{\mathbf{X}_0}{\underline{n}^\top (\mathbf{X} - \mathbf{X}_0)} \left(\frac{\dot{D}}{Z} + \dot{\underline{n}}^\top \mathbf{x} \right) \tag{5.17}$$

Applying the time derivatives of the plane parameters in (5.12) into (5.17) the interaction matrix of a projected point is

$$\mathbf{L_x} = \frac{1}{\Pi_0} \begin{pmatrix} \dfrac{-AX_0}{Z} & \dfrac{-BX_0}{Z} & \dfrac{-CX_0}{Z} & X_0\varepsilon_1 & X_0\varepsilon_2 & X_0\varepsilon_3 \\[2ex] \dfrac{-AY_0}{Z} & \dfrac{-BY_0}{Z} & \dfrac{-CY_0}{Z} & Y_0\varepsilon_1 & Y_0\varepsilon_2 & Y_0\varepsilon_3 \end{pmatrix}$$

(5.18)

$$\Pi_0 = \mathbf{n}^\top(\mathbf{X}_0 - \mathbf{x}Z)$$
$$(\varepsilon_1, \varepsilon_2, \varepsilon_3) = \underline{\mathbf{n}} \times (x, y, 1)$$

(5.19)

note that Π_0 is the distance of the reference point $\mathbf{X}_0$ to the object. With respect to the interaction matrix given by Motyl et al. [Khadraoui et al., 1996], the number of 3D parameters concerning the laser pointer has been reduced from 8 to 3, i.e. X_0, Y_0 and Z. The orientation $\underline{\mathbf{u}}$ of the laser remains implicit in our equations. Concerning the planar object, the number of parameters has been reduced from 4 to 3 since D has been expressed in function of the image coordinates (x, y), the corresponding depth Z, and the normal vector to the planar object $\underline{\mathbf{n}}$.

The rank of $\mathbf{L_x}$ is always equal to 1, which means that the time variation of the x and y coordinates are linked. As already pointed out by Andreff et al. [Andreff et al., 2002], the image point $\mathbf{x}$ moves always along a straight line (hereafter called *epipolar line*). Andreff et al. did not specify the interaction matrix of a projected point, but the interaction matrix of the distance of the point to a certain origin of the epipolar line. Furthermore, the interaction matrix related to this feature was expressed in a frame centred in the laser reference point (similar to $\mathbf{X}_0$), and was expressed in function of the angles defining the normal of the planar object, the angle between the laser pointer and the camera optical axis, and the distance between the camera centre and $\mathbf{X}_0$. The main problem concerning the feature used by Andreff et al. is that a convention must be taken to chose the sign of the distance from $\mathbf{x}$ to the origin of the epipolar line.

5.2.2   Model of the structured light sensor

This section presents the parametric model of the system composed by the camera and the structured light sensor composed of 4 laser pointers. The parameters of the model are the ones appearing in the interaction matrix of every projected point, which are here summarised

- Reference point of each laser pointer $(X_0, Y_0, 0)$.

- Normalised image point coordinates $\mathbf{x} = \mathbf{X}/Z$ of every projected point.

- Depth Z of the projected points.

The reference points $\mathbf{X}_0$ are determined by the actual pose of the laser-cross with respect to the camera frame. The 3D coordinates $\mathbf{X}$ of the projected points can be calculated from $\mathbf{X}_0$, the orientation of the lasers $\underline{\mathbf{u}}$ and the object pose. Concretely, from the equation of the line modelling a laser pointer and the equation of the planar object

$$\begin{cases} \mathbf{X} &= \mathbf{X}_0 + \mu\underline{\mathbf{u}} \\ 0 &= \underline{\mathbf{n}}^\top\mathbf{X} + D \end{cases} \tag{5.20}$$

we can obtain the depth of the projected point

$$Z = -\frac{u_z(\underline{\mathbf{n}}^\top\mathbf{X}_0 + D)}{\underline{\mathbf{n}}^\top\mathbf{u}} + Z_0 \tag{5.21}$$

and the real normalised coordinates of the image point

$$\begin{aligned} x_{real} &= \frac{u_x}{u_z} + \frac{X_0}{Z} \\ y_{real} &= \frac{u_y}{u_z} + \frac{Y_0}{Z} \end{aligned} \tag{5.22}$$

The following subsections present the values of the model parameters under different types of relative poses between the camera and the laser-cross. First of all, the ideal case is presented where the laser-cross frame is perfectly aligned with the camera frame. Afterwards, the parameters of the model are calculated under different types of misalignment between the camera and the laser-cross.

Ideal model

First of all, let us consider that the structured light sensor is perfectly attached to the camera so that the laser-cross frame perfectly coincides with the camera frame. In such a case the model parameters are shown in Table 5.1. The (x, y) and Z parameters have been calculated taking into account that the ideal orientation of the lasers coincides with the optical axis direction so that $^C\underline{\mathbf{u}} = {}^L\underline{\mathbf{u}} = (0, 0, 1)$.

Table 5.1: Ideal model parameters

Laser	X_0	Y_0	x	y	Z
1	0	L	0	L/Z_1	$-(BL+D)/C$
2	$-L$	0	$-L/Z_2$	0	$(AL-D)/C$
3	0	$-L$	0	$-L/Z_3$	$(BL-D)/C$
4	L	0	L/Z_4	0	$-(AL+D)/C$

Model considering laser-cross misalignment

In this case, we are interested in calculating the model parameters when the laser-cross is not perfectly aligned with the camera frame and not perfectly centred in the camera origin. Such a misalignment is represented in Figure 5.4 and it can be modelled according to a frame transformation matrix $^C\mathbf{M}_L$ which passes from points expressed in the laser-cross frame to the camera frame. The model parameters under these conditions are developed in Appendix B.

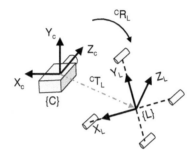

Figure 5.4: Model of misalignment of the laser-cross

5.3 Task definition and stability analysis

5.3.1 Task definition

The goal of our task is to bring the camera to a position where it is parallel to the object. This task corresponds to fixing a plane-to-plane virtual link between the camera image plane and the planar object. Such a virtual link belongs to the class $N = 3$ since this is the number of dof constrained by the link [Espiau *et al.*, 1992]. Concretely, 2 translational and 1 rotational dofs are constrained. This can be seen by stacking the interaction matrices of at least three projected points and evaluating it for $\underline{\mathbf{n}} = (0, 0, 1)$

$$\mathbf{L_x} = \begin{pmatrix} 0 & 0 & X_{01}/Z^2 & y_1 X_{01}/Z & -x_1 X_{01}/Z_1 & 0 \\ 0 & 0 & Y_{01}/Z^2 & y_1 Y_{01}/Z & -x_1 Y_{01}/Z_1 & 0 \\ 0 & 0 & X_{02}/Z^2 & y_2 X_{02}/Z & -x_2 X_{02}/Z_2 & 0 \\ 0 & 0 & Y_{02}/Z^2 & y_2 Y_{02}/Z & -x_2 Y_{02}/Z_2 & 0 \\ 0 & 0 & X_{03}/Z^2 & y_3 X_{03}/Z & -x_3 X_{03}/Z_3 & 0 \\ 0 & 0 & Y_{03}/Z^2 & y_3 Y_{03}/Z & -x_3 Y_{03}/Z_3 & 0 \end{pmatrix} \tag{5.23}$$

The rank of the above matrix is 3 if the points are not collinear. This means that there are three types of camera motion, namely V_x, V_y and Ω_z, which will produce no changes in the image. For the case of a general relative pose camera-object it can also be seen by expressing the above matrix in a frame attached to the object as explained in Appendix C. The interaction matrix expressed in the object frame has the following form (see the appendix for the details)

$$
{}^o\mathbf{L_x} = {}^c\mathbf{L_x} \cdot {}^c\mathbf{T}_o =
\begin{pmatrix}
0 & 0 & -X_{01}/(\Pi_{01}Z_1) & X_{01}\eta_1/\Pi_{01} & X_{01}\xi_1/\Pi_{01} & 0 \\
0 & 0 & -Y_{01}/(\Pi_{01}Z_1) & Y_{01}\eta_1/\Pi_{01} & Y_{01}\xi_1/\Pi_{01} & 0 \\
0 & 0 & -X_{02}/(\Pi_{02}Z_2) & X_{02}\eta_2/\Pi_{02} & X_{02}\xi_2/\Pi_{02} & 0 \\
0 & 0 & -Y_{02}/(\Pi_{02}Z_2) & Y_{02}\eta_2/\Pi_{02} & Y_{02}\xi_2/\Pi_{02} & 0 \\
0 & 0 & -X_{03}/(\Pi_{03}Z_3) & X_{03}\eta_3/\Pi_{03} & X_{03}\xi_3/\Pi_{03} & 0 \\
0 & 0 & -Y_{03}/(\Pi_{03}Z_3) & Y_{03}\eta_3/\Pi_{03} & Y_{03}\xi_3/\Pi_{03} & 0
\end{pmatrix}
\tag{5.24}
$$

with

$$
\begin{aligned}
\eta_i &= \frac{1 - A^2}{C}y_i + \frac{A(Bx_i + ACy_i)}{C(1+C)} \\
\xi_i &= \frac{1 - B^2}{C}x_i + \frac{B(Ay_i + BCx_i)}{C(1+C)}
\end{aligned}
$$

The rank of ${}^o\mathbf{L_x}$ is 3 and the kernel is generated by the following base

$$
\{(1,0,0,0,0,0)\,,\ (0,1,0,0,0,0)\,,\ (0,0,0,0,0,1)\}
\tag{5.25}
$$

As can be seen, for any relative pose camera-object there are three dofs of the planar object which cannot be perceived by the camera.

5.3.2   Stability analysis

One interesting aim in visual servoing is concerned on studying whether the control law (see Section 4.2.2)

$$
\mathbf{v} = -\lambda(\widehat{\mathbf{CL_s}})^+\mathbf{C}(\mathbf{s} - \mathbf{s}^*)
\tag{5.26}
$$

is able to regulate the task function

$$
\mathbf{e} = \mathbf{C}(\mathbf{s} - \mathbf{s}^*)
\tag{5.27}
$$

to the desired state $\mathbf{e}^* = \mathbf{0}$ or not. Assuming that $\mathbf{C}$ is constant the derivative of $\mathbf{e}$ is

$$
\dot{\mathbf{e}} = \mathbf{CL_s}\mathbf{v}
\tag{5.28}
$$

The behaviour of $\mathbf{e}$ is described by the closed-loop equation of the system, which is obtained by plugging the control law (5.26) into (5.28)

$$
\dot{\mathbf{e}} = -\lambda\mathbf{CL_s}(\widehat{\mathbf{CL_s}})^+\mathbf{e}
\tag{5.29}
$$

Note that $\mathbf{L_s}$ is the actual interaction matrix in a certain instant of time t, while $\widehat{\mathbf{L_s}}$ is the value of the model used in the control law. The closed-loop equation will be noted hereafter as

$$\dot{\mathbf{e}} = -\lambda \mathbf{M}(\mathbf{e})\mathbf{e} \qquad (5.30)$$

The aim of the stability analysis is to study if the desired state $\mathbf{e}^* = \mathbf{0}$ is an stable equilibrium point which is reached when time approaches infinity.

Let us remember the following basic definitions:

Equilibrium point: $\mathbf{e}^* = \mathbf{0}$ is said to be an equilibrium point if $\dot{\mathbf{e}} = \mathbf{0}$ when $\mathbf{e} = \mathbf{0}$.

Stability: the stability of an equilibrium point is classically defined in the Lyapunov sense. The equilibrium point in the origin is said to be stable if

$$\forall \epsilon > 0 \; \exists \delta > 0 \; s.t. \; \|\mathbf{e}(0)\| < \delta \Rightarrow \|\mathbf{e}(t)\| < \epsilon, \; \forall t \qquad (5.31)$$

Asymptotic stability: the equilibrium point $\mathbf{e}^* = \mathbf{0}$ is asymptotically stable if it is stable and if it is attracting so that

$$\lim_{t \to \infty} \mathbf{e}(t) = \mathbf{e}^* = \mathbf{0} \qquad (5.32)$$

Hereafter we will focus only on the asymptotic stability since it ensures that the equilibrium is reached. The stability analysis of the control law allows us to determine whether $\mathbf{e}^* = \mathbf{0}$ is reached from any starting point (global asymptotic stability) or only when the initial state is nearby the equilibrium (local asymptotic stability).

If the explicit expression of $\mathbf{e}$ in function of time can be obtained by solving the differential equation (5.30), then it can be checked if the task function zeroes when time approaches infinity. However, in most cases it is not possible to obtain such explicit solution. Alternatively, necessary and sufficient conditions for the local asymptotic stability, and sufficient conditions for the global asymptotic stability are hereafter recalled.

Local asymptotic stability

The local asymptotic stability of the equilibrium point $\mathbf{e}^* = \mathbf{0}$ is analysed by evaluating the closed-loop equation of the system (5.30) around $\mathbf{e}^*$

$$\dot{\mathbf{e}}^* = -\lambda \mathbf{M}(\mathbf{e}^*)\mathbf{e}^* \qquad (5.33)$$

where $\mathbf{M}(\mathbf{e}^*)$ is the product of matrices $\mathbf{CL_s}(\widehat{\mathbf{CL_s}})^+$ evaluated in the desired state. Since $\mathbf{M} = \mathbf{M}(\mathbf{e}^*)$ is a constant matrix, it can be diagonalised as $\mathbf{M} = \mathbf{TDT}^{-1}$ being $\mathbf{T}$ a constant frame transformation and $\mathbf{D}$ a constant diagonal matrix. Noting $\mathbf{f} = \mathbf{T}^{-1}\mathbf{e}^*$ we obtain the following differential equation

$$\dot{\mathbf{f}} = -\lambda \mathbf{D} \mathbf{f} \tag{5.34}$$

having as solution

$$\mathbf{f}(t) = \mathbf{f}(0)\exp^{-\lambda \mathbf{D} t} \tag{5.35}$$

therefore, every component of $\mathbf{f}(t)$ is defined by

$$f_i(t) = f_i(0)\exp^{-\lambda d_i t} \tag{5.36}$$

where d_i is the ith element of the diagonal of $\mathbf{D}$. Every solution $f_i(t)$ will converge to 0 as time approaches infinity

$$\lim_{t\to\infty} f_i(0)\exp^{-\lambda d_i t} = 0 \tag{5.37}$$

if and only if $\mathrm{Re}(d_i) > 0$. Note that if $\mathbf{f}(t) = \mathbf{0}$ when $t \to \infty$ so does $\mathbf{e}(t)$. Therefore, since the elements d_i on the diagonal of $\mathbf{D}$ are nothing but the eigenvalues of $\mathbf{M}$, the necessary and sufficient condition for the system (5.30) to be *locally asymptotically stable* is that the eigenvalues of $\mathbf{M}(\mathbf{e}^*)$ must have all positive real part.

Global asymptotic stability

The global asymptotic stability analysis of non-linear systems is usually done through the Lyapunov indirect method, which provides sufficient conditions. Lyapunov's indirect method states that given a non-linear system of the form

$$\dot{\mathbf{x}} = \mathbf{f}(\mathbf{x}) \tag{5.38}$$

with a unique zero solution at $\mathbf{x} = 0$, a sufficient condition for this equilibrium point to be globally asymptotically stable is the existence of a scalar function $\mathbf{V}(\mathbf{x})$ which must accomplish:

- $\mathbf{V}(\mathbf{0}) = 0$

- $\mathbf{V}(\mathbf{x}) > 0 \; \forall \mathbf{x} \neq \mathbf{0}$

- $\mathbf{V}(\mathbf{x})$ continuous and differentiable

- $\dot{\mathbf{V}}(\mathbf{x}) < 0 \; \forall \mathbf{x}$

if $\mathbf{V}(\mathbf{x})$ accomplishes all this properties then it is said to be a *Lyapunov function*.

Therefore, in order to demonstrate that the task function $\mathbf{e}$ converges to $\mathbf{0}$ for any initial state, it is necessary to find a Lyapunov function $\mathbf{V}(\mathbf{e})$. We assume of course that in the initial state the object is not at the infinity and its orientation is less than $90°$, since then the lasers do not project onto the object and the task function cannot be measured.

First of all, it is necessary to prove that our differential system

$$\dot{\mathbf{e}} = -\lambda \mathbf{M}(\mathbf{e})\mathbf{e} \tag{5.39}$$

has a unique equilibrium point at $\mathbf{e} = \mathbf{0}$. This is true if and only if $\det(\mathbf{M}) \neq 0 \ \forall \mathbf{e}$, since then, the kernel of $\mathbf{M}$ is empty. As in [Malis and Chaumette, 2002], we use the following Lyapunov candidate function

$$\mathbf{V}(\mathbf{e}) = \frac{1}{2}\mathbf{e}^{\top}\mathbf{e} \tag{5.40}$$

Note that this function is always positive and it only zeroes when $\mathbf{e} = \mathbf{0}$. Its derivative is

$$\dot{\mathbf{V}}(\mathbf{e}) = \mathbf{e}^{\top}\dot{\mathbf{e}} = -\lambda \mathbf{e}^{\top}\mathbf{M}\mathbf{e} \tag{5.41}$$

Therefore, if $\mathbf{M}$ is positive definite then $\dot{\mathbf{V}}(\mathbf{e}) < 0$ and $\mathbf{V}(\mathbf{e})$ is a Lyapunov function. Matrix $\mathbf{M}$ is positive definitive if and only if its symmetric part

$$\mathbf{S} = \frac{1}{2}(\mathbf{M} + \mathbf{M}^{\top}) \tag{5.42}$$

has all positive eigenvalues. Therefore, this is a sufficient condition to ensure the global asymptotic stability of the equilibrium point $\mathbf{e}^{*} = \mathbf{0}$.

Note that this sufficient condition ensures that the norm of the task function $\|\mathbf{e}\|$ decreases at every iteration towards 0, since the Lyapunov function that has been chosen can be expressed as

$$\mathbf{V}(\mathbf{e}) = \frac{1}{2}\|\mathbf{e}\|^{2} \tag{5.43}$$

therefore, if $\mathbf{V}(\mathbf{e})$ decreases to 0, so does $\|\mathbf{e}\|$.

5.4 Object plane parameters approach

The first visual servoing approach that we present is a pure position-based method. Indeed, we can use the triangulation capabilities of the system composed by the camera and the lasers in order to reconstruct up to 4 points of the object so that its pose can be recovered. In such case, the $3D$ parameters of the reconstructed plane can be directly used in the closed-loop of the control scheme so that a position-based approach is performed.

Let us consider that the four parameters of the planar object A, B, C and D can be precisely estimated at each iteration. The feature vector $\mathbf{s}$ could be built up by using these four $3D$ parameters. However, since the number of controlled dof is 3, a feature vector of the same dimension is going to be defined. The equation of the object can be noted as a relationship between the depth of a point and its normalised image coordinates as follows

$$\frac{1}{Z} = P_1 + P_2 y + P_3 x \tag{5.44}$$

with $P_1 = -C/D$, $P_2 = -B/D$ and $P_3 = -A/D$. By using the time derivatives of $\underline{\mathbf{n}}$ and D in (5.12) the interaction matrix of $\mathbf{s} = (P_1,\ P_2,\ P_3)$ is calculated obtaining

$$
\mathbf{L_s} = \begin{pmatrix} P_1 P_3 & P_1 P_2 & P_1^2 & -P_2 & P_3 & 0 \\ P_2 P_3 & P_2^2 & P_1 P_2 & P_1 & 0 & -P_3 \\ P_3^2 & P_2 P_3 & P_1 P_3 & 0 & -P_1 & P_2 \end{pmatrix} \tag{5.45}
$$

which in the desired state it has the following value

$$
\mathbf{L_s^*} = \begin{pmatrix} 0 & 0 & 1/Z^{*2} & 0 & 0 & 0 \\ 0 & 0 & 0 & 1/Z^* & 0 & 0 \\ 0 & 0 & 0 & 0 & 0-1/Z^* & 0 \end{pmatrix} \tag{5.46}
$$

On the other hand, the depth of the points belonging to the planar object can be also expressed as

$$
Z = \gamma + \beta Y + \alpha X \tag{5.47}
$$

where $\gamma = -D/C$, $\beta = -B/C$ and $\alpha = -A/C$. In this case, the interaction matrix of the parameters $\mathbf{s} = (\gamma,\ \beta,\ \alpha)$ is

$$
\mathbf{L_s} = \begin{pmatrix} \alpha & \beta & -1 & -\gamma\beta & \gamma\alpha & 0 \\ 0 & 0 & 0 & -1-\beta^2 & \beta\alpha & -\alpha \\ 0 & 0 & 0 & -\beta\alpha & 1+\alpha^2 & \beta \end{pmatrix} \tag{5.48}
$$

Note that in this case the level of decoupling between the 3D features is higher. Furthermore, if we look at the interaction matrix in the desired state

$$
\mathbf{L_s^*} = \begin{pmatrix} 0 & 0 & -1 & 0 & 0 & 0 \\ 0 & 0 & 0 & -1 & 0 & 0 \\ 0 & 0 & 0 & 0 & 1 & 0 \end{pmatrix} \tag{5.49}
$$

we can see that it does not depend on the depth as in the case of $\mathbf{s} = (P_1,\ P_2,\ P_3)$ shown in (5.46). Therefore, in this case the dynamics of the object parameters around the desired state vary linearly with respect to the camera motion. That is why we prefer to use the object plane representation based on $\mathbf{s} = (\gamma,\ \beta,\ \alpha)$. Since the dimension of $\mathbf{s}$ is 3 the control law is

$$
\mathbf{v} = -\lambda \widehat{\mathbf{L_s}}^+ (\mathbf{s} - \mathbf{s}^*) \tag{5.50}
$$

In order to estimate the object plane parameters it is necessary to reconstruct the four 3D points $\mathbf{X}$ projected by the lasers. Then, the equation of the plane best fitting the four points can be calculated by means of least squares. First of all, it is necessary to calculate the 3D point coordinates of every projected laser. The simplest way is to triangulate the points by using the corresponding image normalised coordinates and the laser orientation $\mathbf{u}$ and the laser origin $\mathbf{X_0}$. Nevertheless, it is possible to reduce the number of parameters concerning the laser calibration by using the information provided by the desired image.

Remember that the 3D point $\mathbf{X}$ projected by a certain laser of orientation $\underline{\mathbf{u}}$ and origin $\mathbf{X}_0$ must accomplish the following relationship

$$\mathbf{X} = \mathbf{x}Z = \mu\underline{\mathbf{u}} + \mathbf{X}_0 \tag{5.51}$$

Given an image point $\mathbf{x}^*$ from the desired image the following relationships are extracted from the above equation

$$\begin{cases} x^* Z^* &=& \mu^* u_x + X_0 \\ y^* Z^* &=& \mu^* u_y + Y_0 \\ Z^* &=& \mu^* u_z \end{cases} \tag{5.52}$$

Then, from the last equation we have that $\mu^* = Z^*/u_z$ so that plugging it onto the others we get

$$\begin{aligned} x^* Z^* &= Z^* u_x/u_z + X_0 \\ y^* Z^* &= Z^* u_y/u_z + Y_0 \end{aligned} \tag{5.53}$$

so that the origin of the laser X_0 can be expressed as follows

$$\begin{aligned} X_0 &= Z^*(x^* - u_{xz}) \\ Y_0 &= Z^*(y^* - u_{yz}) \end{aligned} \tag{5.54}$$

where $u_{xz} = u_x/u_z$ and $u_{yz} = u_y/u_z$.

By using the above definitions, the equations in (5.53) can be written for the current image as

$$\begin{aligned} xZ &= Zu_{xz} + Z^*(x^* - u_{xz}) \\ yZ &= Zu_{yz} + Z^*(y^* - u_{yz}) \end{aligned} \tag{5.55}$$

From (5.47) the depth is related to the object parameters as

$$Z = \frac{\gamma}{1 - \alpha x - \beta y} \tag{5.56}$$

so that the equations in (5.55) can be expressed in terms of the object parameters as follows

$$\begin{aligned} \alpha x Z^*(x^* - u_{xz}) + \beta y Z^*(x^* - u_{xz}) + \gamma(x - u_{xz}) - Z^*(x^* - u_{xz}) &= 0 \\ \alpha x Z^*(y^* - u_{yz}) + \beta y Z^*(y^* - u_{yz}) + \gamma(y - u_{yz}) - Z^*(y^* - u_{yz}) &= 0 \end{aligned} \tag{5.57}$$

Then, using these equations for every one of the four laser pointers the following system of non-linear equations is obtained

$$
\left\{
\begin{array}{rcl}
\alpha x_1 Z^*(x_1^* - u_{xz}) + \beta y_1 Z^*(x_1^* - u_{xz}) + \gamma(x_1 - u_{xz}) - Z^*(x_1^* - u_{xz}) &=& 0 \\
\alpha x_1 Z^*(y_1^* - u_{yz}) + \beta y_1 Z^*(y_1^* - u_{yz}) + \gamma(y_1 - u_{yz}) - Z^*(y_1^* - u_{yz}) &=& 0 \\
&\vdots& \\
&=& \vdots \\
\alpha x_4 Z^*(x_4^* - u_{xz}) + \beta y_4 Z^*(x_4^* - u_{xz}) + \gamma(x_4 - u_{xz}) - Z^*(x_4^* - u_{xz}) &=& 0 \\
\alpha x_4 Z^*(y_4^* - u_{yz}) + \beta y_4 Z^*(y_4^* - u_{yz}) + \gamma(y_4 - u_{yz}) - Z^*(y_4^* - u_{yz}) &=& 0
\end{array}
\right.
\tag{5.58}
$$

Note that there are 8 equations for 5 unknowns which are

$$
\xi = (\alpha, \beta, \gamma, \ u_{xz}, \ u_{yz})
\tag{5.59}
$$

Therefore, all the four lasers are assumed to have the same orientation. The system can be numerically solved by a minimisation algorithm based on non-linear least squares like Gauss-Newton or Levenberg-Marquardt. Nevertheless, it cannot be analytically ensured that the algorithm always converges to the right solution. Under calibration errors and image noise, it is possible to reach local minima. Therefore, demonstrating analytically the global asymptotic stability of this position-based approach seems out of reach.

In the following subsection simulations using this position-based approach are presented.

5.4.1   Simulation results

The simulations have been performed by taking into account a sampling time of $\Delta t = 40$ ms and the camera intrinsic parameters obtained from the experimental setup (see Section 5.8). The laser-cross has been simulated using $L = 15$ cm according to the real experimental setup. The desired position has been chosen so that the camera is parallel to the plane at $Z^* = 60$ cm. The initial position the camera is at a distance of 105 cm from the plane and the relative orientation camera-object is defined by $\alpha_x = -30°$ and $\alpha_y = 15°$ according to the specification given in Appendix A. The gain λ has been set to 0.12.

Ideal system

A first simulation has been done by taking into account a perfect alignment of the laser-cross with the camera frame. Furthermore, it has been assumed that the camera intrinsic parameters are perfectly known and all the lasers have the same direction (which coincides in this case with the optical axis direction). The initial and desired image simulated under these conditions are shown in Figure 5.5. As can be seen, the epipolar lines of the lasers $1 - 3$ and $2 - 4$ are perfectly orthogonal and intersect in the central point of the image.

non-constant control law: Figure 5.6 shows the results when $\widehat{\mathbf{L_s}}$ is estimated at each iteration by using the reconstructed object plane parameters. Figure 5.6c shows the co-

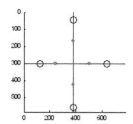

Figure 5.5: Simulation of the ideal system when using $3D$ visual servoing. The initial point distribution is shown with the red dots. The desired point distribution is depicted by the circles. The epipolar lines are painted in blue.

ordinates of a fixed point expressed in the camera frame along the simulation. The fixed point has been set as the initial position of the camera origin. Note that the camera trajectory is almost a straight line in the cartesian space as can also be observed in Figure 5.6d. This is possible since the object pose is perfectly reconstructed under the ideal conditions. Furthermore, the task function has a pure exponential decrease since $\mathbf{L_s} = \mathbf{L_s}$ and the closed-loop equation of the system becomes

$$\dot{\mathbf{e}} = -\lambda \mathbf{e} \tag{5.60}$$

constant control law: the results when using the constant control law based on $\widehat{\mathbf{L_s}} = \mathbf{L_s^*}$ in (5.49) are plotted in Figure 5.7. As can be seen, even if the camera trajectory is no longer almost a straight line, the lateral displacements of the camera are quite small. On the other hand, both the task function components and the camera velocities are strictly monotonic thanks to the linear link existing between them near the desired position (as can be seen in the form of $\mathbf{L_s^*}$).

System including laser-cross misalignment and image noise

A second simulation including calibration errors and image noise has been performed. First, the laser-cross has been displaced from the camera origin according to the translation vector $(4, 10, 9)$ cm. Then, the laser-cross has been rotated $12°$ about its Z axis, $9°$ about Y and $-15°$ about X. The rest of model assumptions still fit (all the lasers have the same relative direction and perfect camera calibration). However, random gaussian noise with standard deviation of 0.5 pixels has been added to the images at each iteration.

The initial position of the camera is still at 105 cm from the object but their relative orientation is defined by $\alpha_x = -25°$ and $\alpha_y = 15°$. The initial and desired image are shown in Figure 5.8. Note that the large misalignment of the laser-cross is evident in these images. However, note that all the epipolar lines intersect in a unique image point. This only happens when all the laser pointers have the same direction.

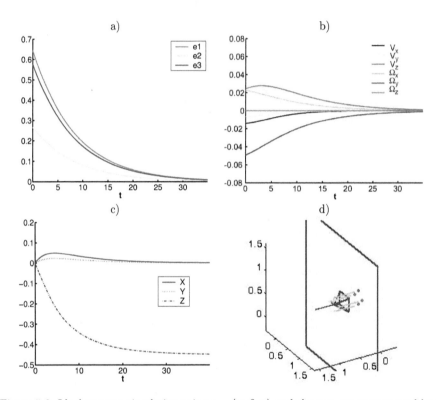

Figure 5.6: Ideal system: simulation using $\mathbf{s} = (\gamma, \beta, \alpha)$ and the non-constant control law. a) $\mathbf{e} = \mathbf{s} - \mathbf{s}^*$ vs. time (in s). b) Camera velocities (ms/s and rad/s) vs. time. c) Fixed point coordinates in the camera frame. d) Scheme of the camera trajectory.

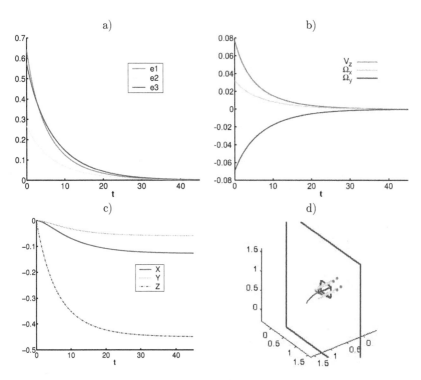

Figure 5.7: Ideal system: simulation using $\mathbf{s} = (\gamma, \beta, \alpha)$ and the constant control law. a) $\mathbf{e} = \mathbf{s} - \mathbf{s}^*$ vs. time (in s). b) Camera velocities (ms/s and rad/s) vs. time. c) Fixed point coordinates in the camera frame. d) Scheme of the camera trajectory.

As can be seen in Figure 5.9 and Figure 5.10 the behaviour of both control laws when using $\mathbf{s} = (\gamma, \beta, \alpha)$ is robust against large misalignment of the laser-cross. The image noise mainly affects the components of the task function e_2 and e_3 while e_1 remains almost insensitive to it.

Remark: the success of the position-based approach in front of large calibration errors relies on the iterative minimisation of the non-linear system of equations which leads to a robust depth estimation of the four projected laser points. During this simulation we have detected certain sensitivity of the numeric algorithm in front of image noise. Therefore, a robust algorithm of minimisation must be used.

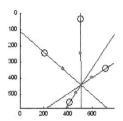

Figure 5.8: Simulation of the system including large laser-cross misalignment. The initial point distribution is shown with the red dots. The desired point distribution is depicted by the circles. The epipolar lines are painted in blue.

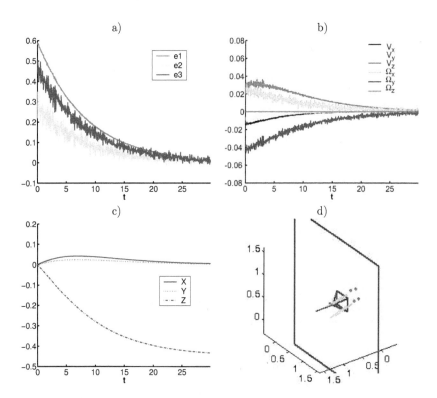

Figure 5.9: System including large laser-cross misalignment and image noise: simulation using $\mathbf{s} = (\gamma, \beta, \alpha)$ and the non-constant control law. a) $\mathbf{e} = \mathbf{s} - \mathbf{s}^*$ vs. time (in s). b) Camera velocities (ms/s and rad/s) vs. time. c) Fixed point coordinates in the camera frame. d) Scheme of the camera trajectory.

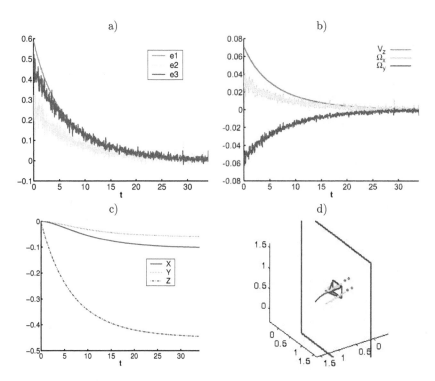

Figure 5.10: System including large laser-cross misalignment and image noise: simulation using $\mathbf{s} = (\gamma, \beta, \alpha)$ and the constant control law. a) $\mathbf{e} = \mathbf{s} - \mathbf{s}^*$ vs. time (in s). b) Camera velocities (ms/s and rad/s) vs. time. c) Fixed point coordinates in the camera frame. d) Scheme of the camera trajectory.

5.5 Image points approach

The simplest $2D$ visual servoing approach that can be defined consists in using the image coordinates of the four projected points. According to the ideal model, if the laser-cross is aligned with the camera frame, the coordinates x_1, y_2, x_3 and y_4 of the four points remain always to 0. Therefore, we can chose as visual features the following vector

$$\mathbf{s} = (y_1, \ x_2, \ y_3, \ x_4) \tag{5.61}$$

Since the number of visual features $k = 4$ is greater than the number of dofs that must be controlled ($m = 3$), matrix $\mathbf{W}$ is chosen so that its rows are the basis of the row space generated by $\widehat{\mathbf{L}_{\mathbf{s}}}$

$$\mathbf{W} = \begin{pmatrix} 0 & 0 & 1 & 0 & 0 & 0 \\ 0 & 0 & 0 & 1 & 0 & 0 \\ 0 & 0 & 0 & 0 & 1 & 0 \end{pmatrix} \tag{5.62}$$

Then, setting $\mathbf{C} = \mathbf{W}\widehat{\mathbf{L}_{\mathbf{s}}}^{+}$, the control law in (5.26) becomes

$$\mathbf{v} = -\lambda \left(\mathbf{W}\widehat{\mathbf{L}_{\mathbf{s}}}^{+}\widehat{\mathbf{L}_{\mathbf{s}}} \right)^{+} \mathbf{W}\widehat{\mathbf{L}_{\mathbf{s}}}^{+} (\mathbf{s} - \mathbf{s}^{*}) \tag{5.63}$$

and the closed-loop equation of the system in (5.29) when using image points is

$$\dot{\mathbf{e}} = -\lambda \mathbf{W}\widehat{\mathbf{L}_{\mathbf{s}}}^{+}\mathbf{L}_{\mathbf{s}} \left(\mathbf{W}\widehat{\mathbf{L}_{\mathbf{s}}}^{+}\widehat{\mathbf{L}_{\mathbf{s}}} \right)^{+} \mathbf{e} \tag{5.64}$$

A constant interaction matrix is used in the control law, which is obtained by using the parameters of the ideal model presented in Table 5.1 (see Section 5.2.2), evaluated in the desired position where $\underline{\mathbf{n}} = (0, 0, 1)$ and $D = -Z^{*}$.

$$\widehat{\mathbf{L}_{\mathbf{s}}} = \mathbf{L}_{\mathbf{s}}^{*} = \begin{pmatrix} 0 & 0 & L/Z^{*2} & L^{2}/Z^{*2} & 0 & 0 \\ 0 & 0 & -L/Z^{*2} & 0 & -L^{2}/Z^{*2} & 0 \\ 0 & 0 & -L/Z^{*2} & L^{2}/Z^{*2} & 0 & 0 \\ 0 & 0 & L/Z^{*2} & 0 & -L^{2}/Z^{*2} & 0 \end{pmatrix} \tag{5.65}$$

In the following subsections three studies of stability are faced. First, we show that the global asymptotic stability of the ideal model cannot be proved. Afterwards, the local asymptotic stability of the system is analysed taking into account certain types of laser-cross misalignment. Finally, the local asymptotic stability in front of errors in the camera intrinsic parameters is also studied.

5.5.1 Global asymptotic stability under perfect conditions

The general expression of $\mathbf{L}_{\mathbf{s}}$ is obtained by using the ideal model parameters in Table 5.1 (see Section 5.2.2) which are expressed in function of the object parameters. The obtained matrix is

$$\mathbf{L}_{\mathbf{s}} = \begin{pmatrix} \frac{LAC}{(BL+D)^{2}} & \frac{LBC}{(BL+D)^{2}} & \frac{LC^{2}}{(BL+D)^{2}} & \frac{L(B^{2}L+BD+LC^{2})}{(BL+D)^{2}} & -\frac{LA}{BL+D} & -\frac{L^{2}AC}{(BL+D)^{2}} \\ -\frac{LAC}{(AL-D)^{2}} & -\frac{LBC}{(AL-D)^{2}} & -\frac{LC^{2}}{(AL-D)^{2}} & \frac{BL}{AL-D} & -\frac{L(LA^{2}+LC^{2}-AD)}{(AL-D)^{2}} & \frac{L^{2}BC}{(AL-D)^{2}} \\ \frac{LAC}{(D-BL)^{2}} & \frac{LBC}{(D-BL)^{2}} & \frac{LC^{2}}{(D-BL)^{2}} & -\frac{L(BD-LB^{2}-LC^{2})}{(D-BL)^{2}} & \frac{LA}{D-BL} & -\frac{L^{2}AC}{(D-BL)^{2}} \\ \frac{LAC}{(AL+D)^{2}} & \frac{LBC}{(AL+D)^{2}} & \frac{LC^{2}}{(AL+D)^{2}} & \frac{BL}{AL+D} & -\frac{L(LA^{2}+LC^{2}+AD)}{(AL+D)^{2}} & \frac{L^{2}BC}{(AL+D)^{2}} \end{pmatrix} \tag{5.66}$$

A sufficient condition for the system to be globally asymptotically stable is that the product of matrices $\mathbf{M}$ in the closed-loop equation is positive definite. The positiveness of $\mathbf{M}$ is ensured if all the eigenvalues of its symmetric part $\mathbf{S}$ are positive.

When using the constant control law based on $\mathbf{L_s^*}$ the analytic expression of the eigenvalues are too complex. On the other hand, if a non-constant law based on estimating $\mathbf{L_s}$ at each iteration is used (reconstructing the object pose parameters involved by triangulation), $\mathbf{M}$ is the identity so that the global asymptotic stability of the ideal model is ensured for any initial object pose.

In presence of calibration errors, the global asymptotic stability analysis becomes too complex. That is why we use instead the local asymptotic stability analysis.

5.5.2 Local asymptotic stability analysis under laser-cross misalignment

The local asymptotic stability analysis is based on studying the real part of the eigenvalues of the matrices product appearing in the closed-loop equation (5.64) evaluated in the desired state

$$\mathbf{M}(\mathbf{e}^*) = \mathbf{W}\widehat{\mathbf{L_s}}^+ \mathbf{L_s}(\mathbf{e}^*) \left(\mathbf{W}\widehat{\mathbf{L_s}}^+ \widehat{\mathbf{L_s}}\right)^+ \tag{5.67}$$

where $\mathbf{L_s}(\mathbf{e}^*)$ is the real interaction matrix at the desired state which takes into account the actual pose of the laser-cross. We take the model parameters presented in Appendix B in order to obtain such an interaction matrix. The study of stability when using an estimation of $\widehat{\mathbf{L_s}}$ at each iteration becomes too complex. The case of using the constant control law based on $\mathbf{L_s^*}$ is considered in the following sections.

Misalignment consisting of a translation

In this case we assume that the cross-laser frame has the same orientation as the camera frame, but that its centre has been displaced according to the vector $^C\mathbf{T}_L = (t_x,\ t_y,\ t_z)$. In this case the parameters in Table B.1 are evaluated for $A = B = 0$, $C = 1$ and $D = -Z^*$ in order to obtain the expression of $\mathbf{L_s}(\mathbf{e}^*)$ which is

$$\mathbf{L_s}(\mathbf{e}^*) = \begin{pmatrix} 0 & 0 & \frac{L+t_y}{Z^{*2}} & \frac{(L+t_y)^2}{Z^{*2}} & -\frac{(L+t_y)t_x}{Z^{*2}} & 0 \\ 0 & 0 & -\frac{L-t_x}{Z^{*2}} & -\frac{t_y(L-t_x)}{Z^{*2}} & -\frac{(L-t_x)^2}{Z^{*2}} & 0 \\ 0 & 0 & -\frac{L-t_y}{Z^{*2}} & \frac{(L-t_y)^2}{Z^{*2}} & \frac{(L-t_y)t_x}{Z^{*2}} & 0 \\ 0 & 0 & \frac{L+t_x}{Z^{*2}} & \frac{t_y(L+t_x)}{Z^{*2}} & -\frac{(L+t_x)^2}{Z^{*2}} & 0 \end{pmatrix} \tag{5.68}$$

The system is locally asymptotically stable if and only if the eigenvalues of $\mathbf{M}(\mathbf{e}^*)$ have all positive real part. The eigenvalues of $\mathbf{M}(\mathbf{e}^*)$ are

$$\sigma_1 = 1$$

$$\sigma_2 = \frac{t_x^2 + t_y^2 + 2L^2 + \sqrt{(t_x^2 + t_y^2)^2 + 6L^2(t_x^2 + t_y^2)}}{2L^2}$$

$$\sigma_3 = \frac{t_x^2 + t_y^2 + 2L^2 - \sqrt{(t_x^2 + t_y^2)^2 + 6L^2(t_x^2 + t_y^2)}}{2L^2}$$

Imposing their positivity the following constraint arises

$$t_x^2 + t_y^2 < 2L^2 \tag{5.69}$$

which means that the local asymptotic stability is only ensured when the projection of the laser-cross centre into the camera plane $Z = 0$ is included in the circle of radius $\sqrt{2}L$ centred in the camera origin (see Figure 5.11 for a schema). Note that the component t_z of the misalignment does not affect the local asymptotic stability. Therefore, a displace-

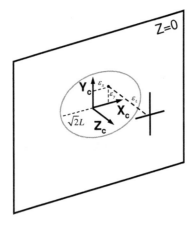

Figure 5.11: Region of local asymptotic stability for the projection of the laser-cross onto the image plane

ment of the laser-cross from the camera origin can strongly affect the global asymptotic stability of the system when using image points since even the local asymptotic stability is constrained.

Misalignment consisting of individual rotations

Let now test the stability of the system when the laser-cross is centred in the camera frame, but rotated with respect to one of the axis. If the three rotations are considered at the same time, too many parameters appear and no analytical results arise.

In the first case, the laser-cross is rotated an angle ψ around the X axis of the camera frame. The interaction matrix $\mathbf{L_s}(\mathbf{e^*})$ is calculated according to the model parameters in Table B.2. The eigenvalues of $\mathbf{M}(\mathbf{e^*})$ are

$$\sigma_1 = 1$$
$$\sigma_2 = \frac{1 + \cos \psi}{2 \cos \psi}$$
$$\sigma_3 = \frac{1}{\cos^2 \psi}$$

it is easy to see that all the eigenvalues are positive if the rotation angle ψ is expressed in the interval $(-\pi/2, \pi/2)$. Note that there is a singularity for $\psi = -\pi/2$ and $\psi = \pi/2$, since in those configurations the lasers do not intersect the object and therefore the servoing is not possible.

When a rotation θ is done around the Y axis the same eigenvalues are obtained. Finally, the eigenvalues corresponding to the case of a rotation ϕ around the Z axis are

$$\sigma_1 = \cos \phi$$
$$\sigma_2 = \cos^2 \phi + \sqrt{\cos^2 \phi (\cos^2 \phi - 1)}$$
$$\sigma_3 = \cos^2 \phi - \sqrt{\cos^2 \phi (\cos^2 \phi - 1)}$$

The positivity of the first eigenvalue imposes that $\phi \in [-\pi/2, \pi/2]$. In the second and the third eigenvalue, the square root is always of a negative number, so that the real part of both eigenvalues is $\cos^2 \phi$, which is always positive.

In conclusion, the approach based on image points is locally asymptotic stable with respect to individual rotations of the laser-cross around the principal axis of the camera frame.

5.5.3 Local asymptotic stability analysis in presence of camera calibration errors

This section presents the local asymptotic stability analysis of the system when the laser-cross is perfectly aligned with the camera frame but the calibrated intrinsic parameters of the camera are not the real ones.

We model the intrinsic parameters of the camera according to the following matrix

$$\mathbf{A} = \begin{pmatrix} f k_u & 0 & u_0 \\ 0 & f k_v & v_0 \\ 0 & 0 & 1 \end{pmatrix} \tag{5.70}$$

where (u_0, v_0) is the principal point in pixels, f the focal distance (in metres), and (k_u, k_v) the conversion factors from metres to pixels for the horizontal and vertical camera axis, respectively. This matrix expresses how the normalised coordinates $\mathbf{x} = \mathbf{X}/Z$ of a 3D point projects onto a certain pixel $\mathbf{x}_p$ of the image as follows

$$\mathbf{x}_p = \mathbf{A}\mathbf{x} \tag{5.71}$$

When only an estimation $\widetilde{\mathbf{A}}$ of the real intrinsic parameters is available, an estimation $\widetilde{\mathbf{x}}$ of the real normalised coordinates $\mathbf{x}$ is obtained from the pixel coordinates

$$\widetilde{\mathbf{x}} = \widetilde{\mathbf{A}}^{-1}\mathbf{x}_p \tag{5.72}$$

This estimation is related to the real normalised coordinates by

$$\widetilde{\mathbf{x}} = \widetilde{\mathbf{A}}^{-1}\mathbf{A}\mathbf{x} \tag{5.73}$$

Hereafter, the elements of $\widetilde{\mathbf{A}}^{-1}\mathbf{A}$ will be noted as follows

$$\widehat{\mathbf{A}}^{-1}\mathbf{A} = \begin{pmatrix} \dfrac{fk_u}{\widetilde{fk_u}} & 0 & \dfrac{u_0 - \widetilde{u}_0}{\widetilde{fk_u}} \\ 0 & \dfrac{fk_v}{\widetilde{fk_v}} & \dfrac{v_0 - \widetilde{v}_0}{\widetilde{fk_v}} \\ 0 & 0 & 1 \end{pmatrix} = \begin{pmatrix} K_u & 0 & U_0 \\ 0 & K_v & V_0 \\ 0 & 0 & 1 \end{pmatrix} \tag{5.74}$$

We assume that $K_u > 0$ and $K_v > 0$ since f, k_u and k_v are positive by definition.

With this notation, the estimated normalised coordinates are related to the real ones as

$$\begin{aligned} \widetilde{x} &= K_u x + U_0 \\ \widetilde{y} &= K_v y + V_0 \end{aligned} \tag{5.75}$$

and therefore, its time derivatives are

$$\begin{aligned} \dot{\widetilde{x}} &= K_u \dot{x} \\ \dot{\widetilde{y}} &= K_v \dot{y} \end{aligned} \tag{5.76}$$

so that

$$\begin{aligned} \mathbf{L}_{\widetilde{x}} &= K_u \mathbf{L}_x \\ \mathbf{L}_{\widetilde{y}} &= K_v \mathbf{L}_y \end{aligned} \tag{5.77}$$

By using the equations above, it is easy to calculate the interaction matrix $\mathbf{L}_{\widetilde{s}}$ corresponding to the visual features set $\widetilde{\mathbf{s}}$ measured under a bad camera calibration. Then, the local asymptotic stability analysis must be applied to the closed-loop equation of the measured task function $\widetilde{\mathbf{e}}$

$$\dot{\widetilde{\mathbf{e}}} = \mathbf{CL}_{\widetilde{\mathbf{s}}}(\widetilde{\mathbf{e}}) \left(\widehat{\mathbf{CL}_{\mathbf{s}}} \right)^{+} \widetilde{\mathbf{e}} \tag{5.78}$$

For the case of $\widetilde{\mathbf{s}} = (\widetilde{y}_1, \widetilde{x}_2, \widetilde{y}_3, \widetilde{x}_4)$ it can be found that the product of matrices $\mathbf{M}(\widetilde{\mathbf{e}})$ in the closed-loop equation of the system evaluated in the desired state becomes

$$\mathbf{M}(\widetilde{\mathbf{e}}^*) = \begin{pmatrix} \dfrac{K_u + K_v}{2} & 0 & 0 \\ 0 & K_v & 0 \\ 0 & 0 & K_u \end{pmatrix} \tag{5.79}$$

whose eigenvalues are in this case the elements of the main diagonal, which are always positive if $K_u > 0$ and $K_v > 0$, which is true if $\widetilde{f} > 0$, $\widetilde{k}_u > 0$ and $\widetilde{k}_v > 0$. Therefore, the local asymptotic stability of the system when using the image point coordinates is ensured if the elements of the main diagonal of $\widetilde{\mathbf{A}}$ are positive.

5.5.4   Simulation results

The system based on the set of visual features $\mathbf{s} = (y_1, \ x_2, \ y_3, \ x_4)$ has been simulated under the same conditions than the ones exposed in Section 5.4.1.

Ideal system

In Figure 5.12 the results of the ideal system when using normalised image points is presented. In this case, the decrease of $\mathbf{s} - \mathbf{s}^*$ is not pure exponential and the rotational velocities generated by the constant control law based on $\mathbf{L}_{\mathbf{s}}^*$ are non-monotonic.

System including laser-cross misalignment and image noise

The system under the calibration errors described in Section 5.4.1 has rapidly diverged when using $\mathbf{s} = (y_1, x_2, y_3, x_4)$. This result was already expected from the local asymptotic stability analysis of this set of visual features when laser-cross misalignment occurs.

5.5.5   Linear combination of image points

As it has been said, in order to fulfill a plane-to-plane virtual link we only need $k = 3$ independent visual features. In the approach based on image points redundant information has been used since the number of visual features was greater than the number of controlled dofs. One might think about linearly combining the image points coordinates in order to obtain a set of 3 visual features.

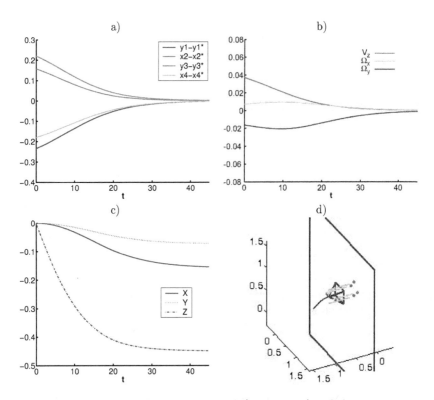

Figure 5.12: Ideal system: simulation using $\mathbf{s} = (y_1, x_2, y_3, x_4)$ and the constant control law. a) $\mathbf{e} = \mathbf{s} - \mathbf{s}^*$ vs. time (in s). b) Camera velocities (ms/s and rad/s) vs. time. c) Fixed point coordinates in the camera frame. d) Scheme of the camera trajectory.

For example, let us define the following set of visual features

$$\mathbf{f} = \left(\begin{array}{ccc} x_4 - x_2 & y_1 + y_3 & x_2 + x_4 \end{array} \right) \tag{5.80}$$

The interaction matrix for the desired position is

$$\mathbf{L_f^*} = \begin{pmatrix} 0 & 0 & 2L/Z^{*2} & 0 & 0 & 0 \\ 0 & 0 & 0 & 2L^2/Z^{*2} & 0 & 0 \\ 0 & 0 & 0 & 0 & -2L^2/Z^{*2} & 0 \end{pmatrix} \tag{5.81}$$

which seems much more decoupled than the interaction matrix corresponding to the image points approach in (5.65).

Let us generalise the definition of the set of 3 visual features $\mathbf{f}$ as a linear combination of $\mathbf{s}$ of the form

$$\mathbf{f} = \mathbf{Qs} \tag{5.82}$$
$$\dot{\mathbf{Q}} = \mathbf{0}$$

so that the interaction matrix in the desired state has the general form

$$\mathbf{L_f^*} = \begin{pmatrix} 0 & 0 & D_1 & 0 & 0 & 0 \\ 0 & 0 & 0 & D_2 & 0 & 0 \\ 0 & 0 & 0 & 0 & D_3 & 0 \end{pmatrix} \tag{5.83}$$

Note that $\mathbf{L_f^*}$ can be decomposed as

$$\mathbf{L_f^*} = \begin{pmatrix} D_1 & 0 & 0 \\ 0 & D_2 & 0 \\ 0 & 0 & D_3 \end{pmatrix} \mathbf{W} = \mathbf{DW} \tag{5.84}$$

On the other hand, by deriving (5.82) we have

$$\dot{\mathbf{f}} = \mathbf{Q}\dot{\mathbf{s}} = \mathbf{QL_s}\mathbf{v} \tag{5.85}$$

so that

$$\mathbf{L_f^*} = \mathbf{QL_s^*} \tag{5.86}$$

Therefore, the following equality holds

$$\mathbf{DW} = \mathbf{QL_s^*} \tag{5.87}$$

and post-multiplying both sides for $\mathbf{L_s^{*+}}$

$$\mathbf{DWL_s^{*+}} = \mathbf{QL_s^*L_s^{*+}} \tag{5.88}$$

since $\mathbf{L_s^*L_s^{*+}} = \mathbf{I}_k$ we have

$$\mathbf{Q} = \mathbf{DWL_s^{*+}} \tag{5.89}$$

The time derivative of the task function when using $\mathbf{s}$ can be expressed as

$$\mathbf{e_s} = \mathbf{C}(\mathbf{s} - \mathbf{s}^*) \Rightarrow \dot{\mathbf{e}}_s = \mathbf{C}\dot{\mathbf{s}} = \mathbf{CL_sv} = \mathbf{WL_s^+L_sv} \tag{5.90}$$

Similarly, for the case of $\mathbf{f}$ we have

$$\mathbf{e_f} = \mathbf{f} - \mathbf{f}^* \Rightarrow \dot{\mathbf{e}}_f = \dot{\mathbf{f}} = \mathbf{L_fv} = \mathbf{QL_s^+L_sv} = \mathbf{DWL_s^+L_sv} \tag{5.91}$$

Note therefore that the dynamics of both tasks are related as follows

$$\dot{\mathbf{e}}_f = \mathbf{D}\dot{\mathbf{e}}_s \tag{5.92}$$

Therefore, the system dynamics of $\mathbf{e_f}$ are identical to the dynamics of $\mathbf{e_s}$ but including a constant factor. Therefore, using a linear combination of visual features which obtains a diagonal interaction matrix in the desired state does not affect the behaviour of the system.

In the following sections new sets of visual features are proposed aiming to improve the performance of the system in terms of stability against calibration errors and decoupling. As it will be seen, the features are based on non-linear combinations of the image points coordinates. Therefore, matrix $\mathbf{Q}$ will depend on the state so that it will be no longer constant and therefore, the dynamics of $\mathbf{e}$ will change.

5.6 Normalised area and angles approach

In this section we analyse the performance of a set of visual features consisting of non-linear combinations of the image points [Pagès *et al.*, 2004].

The first visual feature is based on the area of an element of the image. Such visual feature has been largely used for depth control [Mahony *et al.*, 2002; Tahri and Chaumette, 2004; Corke and Hutchinson, 2001]. In our case, we take into account the area enclosed by the four image points, which can be formulated as follows

$$a = \frac{1}{2}\left((x_3 - x_1)(y_4 - y_2) + (x_2 - x_4)(y_3 - y_1)\right) \tag{5.93}$$

The interaction matrix of the area can be easily derived by using the interaction matrices of the image point coordinates appearing in the formula above. The interaction matrix evaluated in any state where the camera is parallel to the object ($A = 0$, $B = 0$, $C = 1$) at a certain depth Z ($D = -Z$) will be hereafter denoted as $\mathbf{L}^{\|}$. For the case of the area this matrix is

$$\mathbf{L}_a^{\|} = \begin{pmatrix} 0 & 0 & 2a^{\|}/Z & 0 & 0 & 0 \end{pmatrix} \tag{5.94}$$

Note that the area $a^\parallel$ observed in any position where the camera is parallel to the object is known. According to the ideal model it depends on the lasers positions which are symmetrically placed around the camera and pointing towards the same direction than the optical axis. Concretely, we have

$$a^\parallel = \frac{A^\parallel}{Z^2} \tag{5.95}$$

where $A^\parallel$ is the $3D$ area enclosed by the four points onto the object surface whenever the camera is parallel to the object. Since the four laser pointers are orthogonally projected, the $3D$ area is constant for any position where the camera and the object are parallel. Concretely, we have that

$$A^\parallel = 2L^2 \tag{5.96}$$

Therefore, the interaction matrix in (5.94) can be rewritten as

$$\mathbf{L}_a^\parallel = (\ 0\ \ 0\ \ 4L^2/Z^3\ \ 0\ \ 0\ \ 0\ ) \tag{5.97}$$

Note that the dynamics of the area are strongly non-linear.

The 2 visual features controlling the remaining dofs are selected from the 4 virtual segments defined according to Figure 5.13.

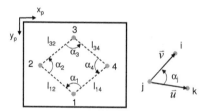

Figure 5.13: At left side, virtual segments defined by the image points. At right side, definition of the angle α_j.

An interesting feature is the angle between each pair of intersecting virtual segments. The angle α_j corresponding to the angle between the segment l_{jk} and the segment l_{ji} (see Figure 5.13) is defined as

$$\sin \alpha_j = \frac{\|\vec{u} \times \vec{v}\|}{\|\vec{u}\|\|\vec{v}\|} \quad , \quad \cos \alpha_j = \frac{\vec{u} \cdot \vec{v}}{\|\vec{u}\|\|\vec{v}\|} \tag{5.98}$$

Then, developing the inner and outer products, the angle is obtained from the point coordinates as follows

$$\alpha_j = \arctan \frac{(x_k - x_j)(y_i - y_j) - (x_i - x_j)(y_k - y_j)}{(x_k - x_j)(x_i - x_j) + (y_k - y_j)(y_i - y_j)} \tag{5.99}$$

Knowing that the derivative of $f(x) = \arctan(x)$ is $\dot{f}(x) = \dot{x}/(1+x^2)$, the interaction matrix of α_j can easily be calculated.

Then, by choosing the visual features $\alpha_{13} = \alpha_1 - \alpha_3$ and $\alpha_{24} = \alpha_2 - \alpha_4$, the following interaction matrices are obtained for the case whenever the camera is parallel to the object

$$
\begin{aligned}
\mathbf{L}_{\alpha_{13}}^{\parallel} &= (\ 0\ \ 0\ \ 0\ \ 2L/Z\ \ \ 0\ \ \ \ 0\ ) \\
\mathbf{L}_{\alpha_{24}}^{\parallel} &= (\ 0\ \ 0\ \ 0\ \ \ 0\ \ \ \ 2L/Z\ \ 0\ )
\end{aligned}
\tag{5.100}
$$

Note that by using the visual feature set $\mathbf{s} = (\ a,\ \alpha_{13},\ \alpha_{24})$ the interaction matrix is diagonal (for any state where the camera and the object are parallel) so that a decoupled control scheme is obtained with no singularities. However, it can be also noted that the non-null terms of the interaction matrix are inversely proportional to the depth Z or a power of the depth Z^3. This will cause the camera trajectory to be not completely satisfactory. As pointed out by Mahony et al. [Mahony et al., 2002], a good visual feature controlling one dof is the one whose error function varies proportionally to the variation of the dof.

Let us start by searching a feature a_n whose time derivative only depends on constant values. Since the time derivative of a depends on the inverse of the depth, we can search a feature of the form $a_n = a^\gamma$ so that the depth is cancelled in its time derivative. Then, taking into account all this, the required power γ can be deduced as follows

$$
a_n = a^\gamma \Rightarrow \dot{a}_n = \gamma a^{\gamma-1}\dot{a} = \frac{2\gamma A^\gamma}{Z^{2\gamma+1}} \cdot V_z
\tag{5.101}
$$

In order to cancel the depth it is necessary that

$$
2\gamma + 1 = 0 \Rightarrow \gamma = -1/2
\tag{5.102}
$$

so that we find $a_n = 1/\sqrt{a}$ as in [Mahony et al., 2002; Tahri and Chaumette, 2003]. The interaction matrix of a_n evaluated in the desired state is in fact valid for any camera position where it is parallel to the object since it only depends on constant values and not on the depth

$$
\mathbf{L}_{a_n}^* = \mathbf{L}_{a_n}^{\parallel} = (\ 0\ \ 0\ \ -1/(\sqrt{2}L)\ \ 0\ \ 0\ \ 0\ )
\tag{5.103}
$$

Following the same method we can find that defining

$$
\begin{aligned}
\alpha_{13n} &= \alpha_{13n}/\sqrt{a} \tag{5.104}\\
\alpha_{24n} &= \alpha_{24n}/\sqrt{a} \tag{5.105}
\end{aligned}
$$

we obtain the following interaction matrices for the new normalised features

$$
\begin{aligned}
\mathbf{L}_{\alpha_{13_n}}^* &= \mathbf{L}_{\alpha_{13_n}}^{\parallel} = (\ 0\ \ 0\ \ 0\ \ \sqrt{2}\ \ \ 0\ \ \ 0\ ) \\
\mathbf{L}_{\alpha_{24_n}}^* &= \mathbf{L}_{\alpha_{24_n}}^{\parallel} = (\ 0\ \ 0\ \ 0\ \ \ 0\ \ \ \sqrt{2}\ \ 0\ )
\end{aligned}
\tag{5.106}
$$

These normalised visual features are related to the object parameters (A, B, C, D) as follows

$$a_n = -\frac{\sqrt{(A^2L^2 - D^2)(B^2L^2 - D^2)}}{\sqrt{2}LCD}$$

$$\alpha_{13n} = a_n \arctan\left(\frac{4LBD(L^2(A^2 + B^2) - 2D^2)}{L^4(A^2 + B^2)^2 + 4D^2(D^2 - 2B^2L^2)}\right)$$

$$\alpha_{24n} = a_n \arctan\left(\frac{4LAD(L^2(A^2 + B^2) - 2D^2)}{L^4(A^2 + B^2)^2 + 4D^2(D^2 - 2A^2L^2)}\right) \qquad (5.107)$$

Similarly, they are related to the object representation $(\gamma = -D/C, \beta = -B/C, \alpha = -A/C)$ by

$$a_n = -\frac{\sqrt{(\alpha^2L^2 - \gamma^2)(\beta^2L^2 - \gamma^2)}}{\sqrt{2}L\gamma}$$

$$\alpha_{13n} = a_n \arctan\left(\frac{4L\beta\gamma(L^2(\alpha^2 + \beta^2) - 2\gamma^2)}{L^4(\alpha^2 + \beta^2)^2 + 4\gamma^2(\gamma^2 - 2\beta^2L^2)}\right)$$

$$\alpha_{24n} = a_n \arctan\left(\frac{4L\alpha\gamma(L^2(\alpha^2 + \beta^2) - 2\gamma^2)}{L^4(\alpha^2 + \beta^2)^2 + 4\gamma^2(\gamma^2 - 2\alpha^2L^2)}\right) \qquad (5.108)$$

If a Taylor approximation of first order is made about $A = B = 0$ or $\alpha = \beta = 0$, the following relationships appear

$$a_n \approx \frac{1}{\sqrt{2}L}\left(\frac{-D}{C}\right) = \frac{1}{\sqrt{2}L}\gamma$$

$$\alpha_{13n} \approx -\sqrt{2}\left(\frac{-B}{C}\right) = -\sqrt{2}\beta \qquad (5.109)$$

$$\alpha_{24n} \approx \sqrt{2}\left(\frac{-A}{C}\right) = \sqrt{2}\alpha$$

Therefore, when the camera is nearly parallel to the object, the features based on normalised area and angles are proportional to the object parameters (γ, β, α). That is why the features $\mathbf{s} = (a_n, \alpha_{13n}, \alpha_{24n})$ are decoupled in the desired state.

Given this set of visual features we have $m = k = 3$ so that $\mathbf{C} = \mathbf{I}_3$ and the control law is

$$\mathbf{v} = -\lambda \widehat{\mathbf{L_s}}^+ (\mathbf{s} - \mathbf{s}^*) \qquad (5.110)$$

and therefore the closed-loop equation of the system is

$$\dot{\mathbf{e}} = -\lambda \mathbf{L_s} \widehat{\mathbf{L_s}}^+ \mathbf{e} \qquad (5.111)$$

When using a constant control law, the estimation of the interaction matrix is simply

$$\widehat{\mathbf{L}_\mathbf{s}} = \mathbf{L}_\mathbf{s}^* = \mathbf{L}_\mathbf{s}^{\|} = \begin{pmatrix} 0 & 0 & -1/(\sqrt{2}L) & 0 & 0 & 0 \\ 0 & 0 & 0 & \sqrt{2} & 0 & 0 \\ 0 & 0 & 0 & 0 & \sqrt{2} & 0 \end{pmatrix} \qquad (5.112)$$

5.6.1 Global asymptotic stability under perfect conditions

Unfortunately, the interaction matrix in function of the object parameters corresponding to the ideal model is very complex. For example, we show the non-null elements of the general interaction matrix for a_n expressed in the object frame (see Appendix C)

$$^{o}\mathbf{L}_{a_n} = \begin{pmatrix} 0 & 0 & ^{o}L_{a_n}(V_z) & ^{o}L_{a_n}(\Omega_x) & ^{o}L_{a_n}(\Omega_y) & 0 \end{pmatrix} \qquad (5.113)$$

$$^{o}L_{a_n}(V_z) = 4\frac{C^2L^2D(A^2B^2L^4 - D^4)}{(A^2L^2 - D^2)^2(B^2L^2 - D^2)^2}$$

$$^{o}L_{a_n}(\Omega_x) = \frac{4L^2CBD^2((1+C)(D^4 - L^2D^2) + A^2L^4(1 - A^2 + C(1+B^2)))}{(1+C)(A^2L^2 - D^2)^2(B^2L^2 - D^2)^2}$$

$$^{o}L_{a_n}(\Omega_y) = -\frac{4L^2CAD^2((1+C)(D^4 - L^2D^2) + B^2L^4(1 - B^2 + C(1 - A^2)))}{(1+C)(A^2L^2 - D^2)^2(B^2L^2 + D^2)^2}$$

The interaction matrices for α_{13n} and α_{24n} are still more complicated because of the definition of their time derivatives

$$\dot{\alpha}_{13n} = a_n \cdot \dot{\alpha}_{13} + \alpha_{13} \cdot \dot{a}_n$$
$$\dot{\alpha}_{24n} = a_n \cdot \dot{\alpha}_{24} + \alpha_{24} \cdot \dot{a}_n \qquad (5.114)$$

Note that when the camera is not parallel to the object α_{13} and α_{24} are different to 0. Then, the general interaction matrices depend on arctan functions.

Trying to analyse the global asymptotic stability of the system when using the constant control law based on matrix (5.112) becomes too complex. We could only ensure the global asymptotic stability of the ideal system when using a non-constant control law based on perfectly estimating $\widehat{\mathbf{L}_\mathbf{s}} = \mathbf{L}_\mathbf{s}$ at each iteration. In this case, the closed-loop equation of the system is

$$\dot{\mathbf{e}} = -\lambda\mathbf{e} \qquad (5.115)$$

obtaining a pure exponential decrease of the task function.

Hereafter we focus on the image-based approach based on $\mathbf{s} = (a_n,\ \alpha_{13n},\ \alpha_{24n})$ and the constant control law. We show its robustness against to calibration errors through the local asymptotic stability analysis.

5.6.2 Local asymptotic stability analysis under laser-cross misalignment

Misalignment consisting of a translation

The interaction matrix for the desired position including a misalignment of the laser-cross consisting of a translation ${}^{C}\mathbf{T}_L = (t_x,\ t_y,\ t_z)$ has been computed taking into account the model parameters in Table B.1 at Appendix B obtaining

$$\mathbf{L_s}(\mathbf{e}^*) = \begin{pmatrix} 0 & 0 & -\frac{\sqrt{2}}{2L} & -\frac{3\sqrt{2}t_y}{4L} & \frac{3\sqrt{2}t_x}{4L} & 0 \\ 0 & 0 & 0 & \sqrt{2} & 0 & 0 \\ 0 & 0 & 0 & 0 & \sqrt{2} & 0 \end{pmatrix} \tag{5.116}$$

Note that the misalignment parameters only affect the normalised area a_n. On the other hand, α_{13n} and α_{24n} are invariant to such type of misalignment near the desired state.

Then, the product of matrices in the closed-loop equation (5.111) becomes $\mathbf{M}(\mathbf{e}^*) = \mathbf{L_s}(\mathbf{e}^*)\widehat{\mathbf{L_s}}^+$ that is

$$\mathbf{M}(\mathbf{e}^*) = \begin{pmatrix} 1 & -\frac{3t_y}{4L} & \frac{3t_x}{4L} \\ 0 & 1 & 0 \\ 0 & 0 & 1 \end{pmatrix} \tag{5.117}$$

Note that the eigenvalues of $\mathbf{M}(\mathbf{e}^*)$ are equal to the elements of the main diagonal. Therefore, the eigenvalues are all equal to 1. It means that the local asymptotic stability of the system when using this set of visual features is not affected by a misalignment of the laser-cross consisting of a translation. Therefore, the stability domain of this set of visual features is much larger than the approach based on image points.

Misalignment consisting of individual rotations

Let us now consider how does a rotation of the laser-cross around one of the axis of the camera frame affect the local asymptotic stability of the system.

Given the case of a rotation ψ around the X axis the eigenvalues of $\mathbf{M}(\mathbf{e}^*)$ are

$$\sigma_1 = \frac{2\cos\psi\sqrt{1/\cos\psi}}{\cos^2\psi + 1}$$

$$\sigma_2 = \frac{2\cos^2\psi\sqrt{1/\cos\psi}}{\cos^2\psi + 1}$$

$$\sigma_3 = \cos\psi\sqrt{1/\cos\psi}$$

which are all positive and definite if $\psi \in (-\pi/2,\ \pi/2)$.

The same eigenvalues are obtained for the case of a rotation θ around the Y axis of the camera.

Finally, when a rotation ϕ is applied around the Z axis, the eigenvalues of $\mathbf{M}(\mathbf{e}^*)$ are

$$
\begin{aligned}
\sigma_1 &= 1 \\
\sigma_2 &= \cos\phi + \sqrt{\cos^2\phi - 1} \\
\sigma_3 &= \cos\phi - \sqrt{\cos^2\phi - 1}
\end{aligned}
$$

Note that the real part of the two last eigenvalues is $\cos\phi$ so in order to ensure their positivity we must have $\phi \in [-\pi/2, \pi/2]$.

It can be stated that the local asymptotic stability of the approach based on the normalised area and angles is not affected by individual rotations of the laser-cross around the camera axis.

5.6.3   Local asymptotic stability analysis in presence of camera calibration errors

We now present the local asymptotic stability analysis taking into account in the real interaction matrix both the real and the estimated intrinsic parameters of the camera. As explained in Section 5.5.3, it is necessary to study the dynamic behaviour of the measured task function $\widetilde{\mathbf{e}}$ evaluated in the desired state

$$
\dot{\widetilde{\mathbf{e}}}^* = -\lambda \mathbf{L}_{\widetilde{\mathbf{s}}}(\widetilde{\mathbf{e}}^*)\widehat{\mathbf{L}}_{\mathbf{s}}^{+}\widetilde{\mathbf{e}}^* \tag{5.118}
$$

In this case, the interaction matrix in the desired state $\widetilde{\mathbf{e}}^*$ is

$$
\mathbf{L}_{\widetilde{\mathbf{s}}}(\widetilde{\mathbf{e}}^*) = \begin{pmatrix}
0 & 0 & -\dfrac{\sqrt{2}(K_u + K_v)}{4L} & 0 & 0 & 0 \\
0 & 0 & 0 & \dfrac{2LK_v}{\sqrt{K_u}Z^*} & 0 & 0 \\
0 & 0 & 0 & 0 & \dfrac{2\sqrt{K_u}L}{Z^*} & 0
\end{pmatrix} \tag{5.119}
$$

so that the product of matrices $\mathbf{M}(\widetilde{\mathbf{e}}^*) = \mathbf{L}_{\mathbf{s}}(\widetilde{\mathbf{e}}^*)\widehat{\mathbf{L}}_{\mathbf{s}}^{+}$ is

$$
\mathbf{M}(\widetilde{\mathbf{e}}^*) = \begin{pmatrix}
\dfrac{K_u + K_v}{2} & 0 & 0 \\
0 & \dfrac{\sqrt{2}LK_v}{\sqrt{K_u}Z^*} & 0 \\
0 & 0 & \dfrac{\sqrt{2K_u}L}{Z^*}
\end{pmatrix} \tag{5.120}
$$

Note that the eigenvalues are the elements of the main diagonal and are positive if $K_u > 0$ and $K_v > 0$, which is true if $\widetilde{f} > 0$, $\widetilde{k}_u > 0$ and $\widetilde{k}_v > 0$.

Therefore, the system based on the normalised area and angles is robust against camera calibration errors if the elements of the main diagonal of $\widetilde{\mathbf{A}}$ are positive.

5.6.4 Simulation results

The simulations described in Section 5.4.1 have been also done taking into account the set of visual features $\mathbf{s} = (a_n, \alpha_{13n}, \alpha_{24n})$ and the constant control law based on $\mathbf{L_s^*}$.

Ideal system

The results obtained according to the ideal system specification are shown in Figure3 5.14. The decrease of the task function is strictly monotonic as well as the camera velocities.

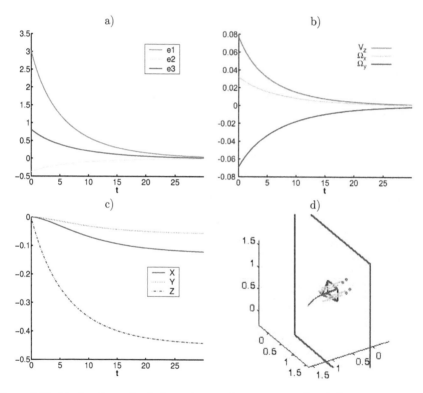

Figure 5.14: Ideal system: simulation using $\mathbf{s} = (a_n, \alpha_{13n}, \alpha_{24n})$ and the constant control law. a) $\mathbf{e} = \mathbf{s} - \mathbf{s}^*$ vs. time (in s). b) Camera velocities (ms/s and rad/s) vs. time. c) Fixed point coordinates in the camera frame. d) Scheme of the camera trajectory.

Note that the camera velocities and the camera trajectory are pretty similar to the ones obtained by the position-based approach in Section 5.4.1 when using the constant control law. This similarity was expected from the analytic approximate behaviour deduced in (5.109).

System including laser-cross misalignment and image noise

Figure 5.15 presents the results obtained by the current image-based approach for the simulation conditions described in Section 5.4.1. Note that the system is almost unaffected by the large laser-cross misalignment, as already expected from the local asymptotic stability analysis results in presence of this type of calibration errors. Again, the results are very similar to the ones obtained by the positioned-based approach when using the constant control law.

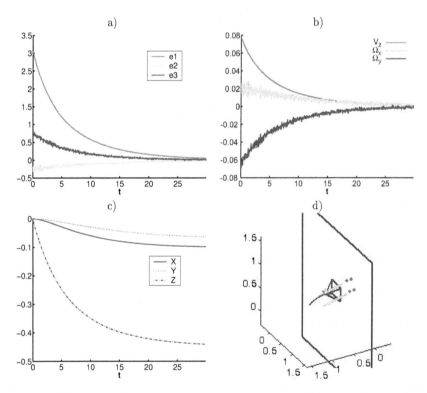

Figure 5.15: System including large laser-cross misalignment and image noise: simulation using $\mathbf{s} = (a_n,\ \alpha_{13n},\ \alpha_{24n})$ and the constant control law. a) $\mathbf{e} = \mathbf{s} - \mathbf{s}^*$ vs. time (in s). b) Camera velocities (ms/s and rad/s) vs. time. c) Fixed point coordinates in the camera frame. d) Scheme of the camera trajectory.

5.7 A decoupled image-based approach

The aim of this section is to obtain a set of 3 visual features which decouples as much as possible the controlled dof not only near the desired position (as in the case of the normalised area and angles) but for any camera-object pose. With such a decoupling we also aim to demonstrate the global asymptotic stability of the system when using the constant control law based on $\mathbf{L_s^*}$. Moreover, we expect a set of visual features which have the same robustness against calibration errors demonstrated for the previous features.

Let us take a look at the interaction matrices of y_1^{-1}, y_3^{-1}, x_2^{-1} and x_4^{-1}

$$\mathbf{L}_{y_1^{-1}} = \left(\begin{array}{cccccc} -K_1 & -K_2 & -K_3 & -\frac{K_2(BL+D)}{C} - 1 & \frac{K_1(BL+D)}{C} & K_1L \end{array} \right)$$

$$\mathbf{L}_{y_3^{-1}} = \left(\begin{array}{cccccc} K_1 & K_2 & K_3 & -\frac{K_2(BL-D)}{C} - 1 & \frac{K_1(BL-D)}{C} & K_1L \end{array} \right)$$

$$\mathbf{L}_{x_2^{-1}} = \left(\begin{array}{cccccc} K_1 & K_2 & K_3 & -\frac{K_2(AL-D)}{C} & \frac{K_1(AL-D)}{C} + 1 & -K_2L \end{array} \right)$$

$$\mathbf{L}_{x_4^{-1}} = \left(\begin{array}{cccccc} -K_1 & -K_2 & -K_3 & -\frac{K_2(AL+D)}{C} & \frac{K_1(AL+D)}{C} + 1 & -K_2L \end{array} \right)$$

with $$K_1 = \frac{A}{LC} \quad K_2 = \frac{B}{LC} \quad K_3 = \frac{1}{L}$$

It is obvious that simple combinations of such features can lead to a decoupled system. We have chosen the following set of visual features

$$\mathbf{s} = \left(\begin{array}{ccc} y_1^{-1} - y_3^{-1}, & y_1^{-1} + y_3^{-1}, & x_2^{-1} + x_4^{-1} \end{array} \right)$$

$$= \left(\begin{array}{ccc} \dfrac{y_3 - y_1}{y_1 y_3}, & \dfrac{y_1 + y_3}{y_1 y_3}, & \dfrac{x_2 + x_4}{x_2 x_4} \end{array} \right) \tag{5.121}$$

whose interaction matrix is

$$\mathbf{L_s} = \left(\begin{array}{cccccc} -\frac{2A}{LC} & -\frac{2B}{LC} & -\frac{2}{L} & -\frac{2BD}{LC^2} & \frac{2AD}{LC^2} & 0 \\ 0 & 0 & 0 & -\frac{2(1-A^2)}{C^2} & \frac{2AB}{C^2} & \frac{2A}{C} \\ 0 & 0 & 0 & -\frac{2AB}{C^2} & \frac{2(1-B^2)}{C^2} & -\frac{2B}{C} \end{array} \right) \tag{5.122}$$

which is always rank 3 unless for degenerated cases. Note that the rotational part is decoupled from the translational one for any camera-object pose. To our knowledge, there are no other image-based approach where the interaction matrix has such a high order of decoupling for any state.

In the ideal case, that is, when no calibration errors occur, the visual features are related to the object parameters as follows

$$\mathbf{s} = (s_1, s_2, s_3) = \left(-\frac{2D}{LC}, -2\frac{B}{C}, -2\frac{A}{C} \right) \tag{5.123}$$

Therefore, these features are proportional to the parameters of the object plane expressed as

$$Z = \gamma + \beta Y + \alpha X \tag{5.124}$$

so that

$$s_1 = \frac{2}{L}\gamma, \ s_2 = 2\beta, \ s_3 = 2\alpha \tag{5.125}$$

Furthermore, taking into account the Taylor developments in (5.109), the new visual features are related to $(a_n, \alpha_{13n}, \alpha_{24n})$ by

$$s_1 \approx 2\sqrt{2}a_n, \ s_2 \approx -\sqrt{2}\alpha_{13n}, \ s_3 \approx \sqrt{2}\alpha_{24n} \tag{5.126}$$

Therefore, under ideal conditions, the image-based approach based on these features behaves as the position-based technique presented in Section 5.4 and very similar to the image-based approach based on $(a_n, \alpha_{13n}, \alpha_{24n})$. Hence, a new way to implicitly estimate the object pose has been found from a non-linear combination of the image point coordinates. In absence of calibration errors the equations in (5.125) could be used to obtain the object parameters (γ, β, α) and execute the position-based approach without need to solve the system of non-linear equations in (5.58).

Another interesting characteristic of these features is that the interaction matrix can be expressed in terms of the task function components. Usually, in most part of $2D$ visual servoing approaches this is unfeasible. By using the normalised image points coordinates for the ideal case (presented in Table 5.1), the components of $\mathbf{e} = \mathbf{s} - \mathbf{s}^* = (e_1, e_2, e_3)$ can be expressed as follows

$$e_1 = \frac{-2(D + CZ^*)}{LC}, \ e_2 = -2\frac{B}{C}, \ e_3 = -2\frac{A}{C} \tag{5.127}$$

We remember that $\underline{\mathbf{n}}$ is a unitary vector so that $C = \sqrt{1 - A^2 - B^2}$. Hence, we have a system of 3 equations and 3 unknowns $(A, B$ and $D)$ whose unique solution is

$$A = -\frac{e_3}{h}, \ B = -\frac{e_2}{h}, \ D = -\frac{e_1 L + 2Z^*}{h} \tag{5.128}$$

with

$$h = \sqrt{e_2^2 + e_3^2 + 4} \tag{5.129}$$

and therefore

$$C = \frac{2}{h} \tag{5.130}$$

Using these equivalences, the interaction matrix can be expressed in terms of the task function components as follows

$$
\mathbf{L_s}(\mathbf{e}) = \begin{pmatrix} \frac{e_3}{L} & \frac{e_2}{L} & -\frac{2}{L} & -\frac{e_2(e_1 L + 2Z^*)}{2L} & \frac{e_3(e_1 L + 2Z^*)}{2L} & 0 \\ 0 & 0 & 0 & -\frac{1}{2}e_2^2 - 2 & \frac{1}{2}e_2 e_3 & -e_3 \\ 0 & 0 & 0 & -\frac{1}{2}e_2 e_3 & 2 + \frac{1}{2}e_3^2 & e_2 \end{pmatrix}
\tag{5.131}
$$

Note that all the terms in the interaction matrix are known. This allows us to decide which model of interaction matrix is used in the control law obtained by setting $\mathbf{W} = \widehat{\mathbf{L_s}}$ so that $\mathbf{C} = \mathbf{I}_3$ and

$$
\mathbf{v} = -\lambda \widehat{\mathbf{L_s}}^+ \mathbf{e}
\tag{5.132}
$$

- $\widehat{\mathbf{L_s}}$ estimated at each iteration. Note that in this case the elements of the interaction matrix can be obtained from the task function, without reconstructing the object parameters by triangulation. The main advantage of such a control law is that if the interaction matrix is perfectly estimated the task function will have a pure exponential decrease. However, the camera velocities may be inadequate due to the non-linearities visible in (5.131).

- $\widehat{\mathbf{L_s}} = \mathbf{L_s^*}$ being

$$
\widehat{\mathbf{L_s}} = \mathbf{L_s^*} = \begin{pmatrix} 0 & 0 & -2/L & 0 & 0 & 0 \\ 0 & 0 & 0 & -2 & 0 & 0 \\ 0 & 0 & 0 & 0 & 2 & 0 \end{pmatrix}
\tag{5.133}
$$

In this case, the control law becomes simpler and can be calculated faster (it is not required to calculate the pseudoinverse at each iteration). Note that this matrix does not contain any non-linearities neither depth information, like in the set of visual features based on the normalised area and angles. In this case however, this result has been achieved without need to normalise the features as before. Then, since each visual feature varies proportionally to the dof which controls, if the task function has a good decreasing, which will be studied in the following subsection, suitable camera velocities will be produced [Mahony et al., 2002].

5.7.1   Global asymptotic stability under perfect conditions

The closed-loop equation of the system is again

$$
\dot{\mathbf{e}} = -\lambda \mathbf{L_s}(\mathbf{e}) \widehat{\mathbf{L_s}}^+ \mathbf{e}
\tag{5.134}
$$

We now present the global asymptotic stability analysis when using two different control laws.

Non-constant control law

In this case we compute the real interaction matrix at each iteration from the task function value so that $\widehat{\mathbf{L}_s} = \mathbf{L}_s(\mathbf{e})$. Then the product of matrices $\mathbf{M}$ in the control law is the identity. Therefore, the equilibrium point is unique and the closed-loop equation becomes simply

$$\dot{\mathbf{e}} = -\lambda \mathbf{e} \tag{5.135}$$

which ensures a pure exponential decrease of the task function. However, the camera velocities produced by the non-constant control law can be not very suitable due to the strong non-linearities in $\mathbf{L}_s(\mathbf{e})^+$.

Constant control law

When using the constant diagonal matrix in (5.133) in the control law, the product of matrices $\mathbf{M} = \mathbf{L}_s\widehat{\mathbf{L}_s}^+ = \mathbf{L}_s\mathbf{L}_s^{*+}$ is the following 3×3 matrix

$$\mathbf{M} = \begin{pmatrix} 1 & \frac{BD}{LC^2} & \frac{AD}{LC^2} \\ 0 & \frac{B^2+C^2}{C^2} & \frac{AB}{C^2} \\ 0 & \frac{AB}{C^2} & \frac{A^2+C^2}{C^2} \end{pmatrix} = \begin{pmatrix} 1 & \frac{e_2(e_1 L+2Z^*)}{4L} & \frac{e_3(e_1 L+2Z^*)}{4L} \\ 0 & \frac{e_2^2}{4}+1 & \frac{e_2 e_3}{4} \\ 0 & \frac{e_2 e_3}{4} & \frac{e_3^2}{4}+1 \end{pmatrix} \tag{5.136}$$

whose determinant is

$$\det(\mathbf{M}) = \frac{1}{C^2} = \frac{h^2}{4} = \frac{\sqrt{e_2^2 + e_3^2 + 4}}{4} \tag{5.137}$$

which is always non-null, and therefore, the equilibrium point $\mathbf{e} = 0$ is unique. The global asymptotic stability analysis can be done by using the Lyapunov method, but only sufficient conditions are provided. However, thanks to the nice decoupled form of the interaction matrix, we can solve the differential system in function of time corresponding to the closed-loop equation of the system

$$\dot{\mathbf{e}}(t) = -\lambda \mathbf{L}_s(\mathbf{e}(t))\widehat{\mathbf{L}_s}^+ \mathbf{e}(t) \tag{5.138}$$

This differential system can be decomposed as follows

$$\dot{e}_1(t) = -\frac{\lambda}{4L}\left(e_1(t)\left(4L + e_2(t)^2 L + e_3(t)^2 L\right) + 2Z^*\left(e_2(t)^2 + e_3(t)^2\right)\right) \tag{5.139}$$

$$\dot{e}_2(t) = -\frac{\lambda}{4}(e_2(t)^3 + 4e_2(t) + e_2(t)e_3(t)^2) \tag{5.140}$$

$$\dot{e}_3(t) = -\frac{\lambda}{4}(e_3(t)^3 + 4e_3(t) + e_3(t)e_2(t)^2) \tag{5.141}$$

The following solutions are obtained according to the developments presented in Appendix D.1

$$e_1(t) = \frac{2e_1(0)}{a(t)} - \frac{2bZ^* \arctan(u(t))}{a(t)L} \tag{5.142}$$

$$e_2(t) = \frac{2e_2(0)}{a(t)} \tag{5.143}$$

$$e_3(t) = \frac{2e_3(0)}{a(t)} \tag{5.144}$$

with

$$a(t) = \sqrt{(e_2^2(0) + e_3^2(0))(\exp^{2\lambda t} - 1) + 4\exp^{2\lambda t}} \tag{5.145}$$

$$b = \sqrt{e_2^2(0) + e_3^2(0)} \tag{5.146}$$

$$u(t) = \frac{b(a(t) - 2)}{b^2 + 2a(t)} \tag{5.147}$$

Let us start by demonstrating the global asymptotic stability of the rotational subsystem defined by (5.140) and (5.141). The subsystem formed by $e_2(t)$ and $e_3(t)$ is globally asymptotically stable if

$$\lim_{t \to \infty} e_2(t) = 0, \quad \lim_{t \to \infty} e_3(t) = 0 \tag{5.148}$$

Both functions clearly tend to 0 when time approaches infinity since $\lim_{t \to \infty} a(t) = \infty$. Moreover, it is easy to show that $e_2(t)$ and $e_3(t)$ are strictly monotonic functions by taking a look at their first derivative

$$\dot{e}_i(t) = -\frac{2\lambda e_i(0) \exp^{2\lambda t} (e_2^2(0) + e_3^2(0) + 4)}{a(t)^3} \tag{5.149}$$

with $i = \{2, 3\}$. Note that the functions e_2 and e_3 are monotonic since the sign of their derivatives never changes and it only depends on the initial conditions. Furthermore, they are strictly monotonic since their derivative only zeroes when $t \to \infty$ or when the function at $t = 0$ is already 0. Therefore, for any initial condition, $e_2(t)$ and $e_3(t)$ always tend towards 0 strictly monotonically.

The global asymptotic stability of the translational subsystem depends on the behaviour of $e_1(t)$. It is easy to show that $e_1(t)$ converges to 0 for any initial state since

$$\left. \begin{array}{rcl} \lim_{t \to \infty} u(t) &=& \frac{b}{2} \\ \lim_{t \to \infty} a(t) &=& \infty \end{array} \right\} \Rightarrow \lim_{t \to \infty} e_1(t) = 0 \tag{5.150}$$

The monotonicity of $e_1(t)$ is not so easy to proof. In fact, depending on the initial conditions, $e_1(t)$ can be not monotonic showing some extrema. In Appendix D.2 it is shown that $e_1(t)$ either is always monotonic or it has a unique extremum before converging monotonically to 0. Furthermore, sufficient conditions are given so that it is possible to check from the initial state of the system and the desired depth Z^* if either $e_1(t)$ will be monotonic during all the servoing or it will have a peak.

5.7.2   Camera trajectory

Thanks to the decoupled form of the interaction matrix in (5.122) we have obtained the analytic functions describing the behaviour of the task function $\mathbf{e}(t)$. Furthermore, when using the constant control law based on $\mathbf{L}_s^*$ we can also obtain the functions describing the camera trajectory. In this case, the control law maps the task function components $e_1(t)$, $e_2(t)$ and $e_3(t)$ to the camera velocities as follows

$$\mathbf{v} = -\lambda \mathbf{L}_s^{*+} \mathbf{e} \tag{5.151}$$

where

$$\mathbf{L}_s^{*+} = \begin{pmatrix} 0 & 0 & 0 \\ 0 & 0 & 0 \\ -L/2 & 0 & 0 \\ 0 & -1/2 & 0 \\ 0 & 0 & 1/2 \\ 0 & 0 & 0 \end{pmatrix} \tag{5.152}$$

so that

$$\begin{cases} V_z(t) &= \lambda \dfrac{L}{2} e_1(t) \\[2mm] \Omega_x(t) &= \lambda \dfrac{1}{2} e_2(t) \\[2mm] \Omega_y(t) &= -\lambda \dfrac{1}{2} e_3(t) \end{cases} \tag{5.153}$$

Then, we can express the coordinates of a fixed point $\mathbf{X}$ in the camera frame in any instant of time t when the camera moves according to $\mathbf{v}(t) = (\mathbf{V}(t)\ \mathbf{\Omega}(t))$ by using the well-known kinematic equation

$$\dot{\mathbf{X}}(t) = -\mathbf{V}(t) - \mathbf{\Omega}(t) \times \mathbf{X}(t) \tag{5.154}$$

Since the constant control law only generates velocities for V_z, Ω_x and Ω_y, the above equation can be rewritten as

$$\begin{cases} \dot{X}(t) &= -\Omega_y(t)Z(t) \\ \dot{Y}(t) &= +\Omega_x(t)Z(t) \\ \dot{Z}(t) &= -V_z(t) + \Omega_y(t)X(t) - \Omega_x(t)Y(t) \end{cases} \tag{5.155}$$

where $V_z(t)$, $\Omega_x(t)$ and $\Omega_y(t)$ are the expressions in (5.153). If we choose as fixed point the initial position of the camera ($\mathbf{X}(0) = (0,0,0)$), we can solve the system of differential equations obtaining

$$
\begin{cases}
X(t) = \dfrac{e_3(0)\exp^{-\lambda t}}{h(0)^2 b^3}\left(\exp^{\lambda t} b^2 Z^* h(0)^2 \arctan\left(u(t)\right) - e_1(0)Lb\left(\exp^{\lambda t} h(0)^2 - b^2 - 2a(t)\right)\right. \\
\qquad\qquad \left. + b^3 Z^*\left(a(t) - 2\right)\right) \\[8pt]
Y(t) = \dfrac{e_2(0)\exp^{-\lambda t}}{h(0)^2 b^3}\left(\exp^{\lambda t} b^2 Z^* h(0)^2 \arctan\left(u(t)\right) - e_1(0)Lb\left(\exp^{\lambda t} h(0)^2 - b^2 - 2a(t)\right)\right. \\
\qquad\qquad \left. + b^3 Z^*\left(a(t) - 2\right)\right) \\[8pt]
Z(t) = \dfrac{-\exp^{-\lambda t}}{h(0)^2}\left(-b^2 Z^*\left(\exp^{\lambda t} - 1\right) + e_1(0)L\left(a(t) - 2\right)2Z^*\left(a(t) - 2\exp^{\lambda t}\right)\right)
\end{cases}
$$

$$(5.156)$$

Note that $\mathbf{X}(t)$ describes how the camera moves farther from its initial position. The expressions of $X(t)$ and $Y(t)$ have the same form, the only difference is that $X(t)$ depends on $e_3(0)$ while $Y(t)$ does on $e_2(0)$. The study of the derivative of $X(t)$ (and similarly for $Y(t)$) shows that both $X(t)$ and $Y(t)$ are monotonic functions. The demonstration is as follows. Let us look at for example at the derivative of $X(t)$

$$
\dot{X}(t) = \frac{\lambda e_3(0)\exp^{\lambda t}}{h(0)^2 a(t)}\left(b^2 Z^*\left(\exp^{\lambda t} - 1\right) + 2Z^*\left(2\exp^{\lambda t} - a(t)\right) + e_1(0)L\left(2 - a(t)\right)\right)
$$

$$(5.157)$$

noting that the sign depends on $e_3(0)$ and will not change if

$$
b^2 Z^*\left(\exp^{\lambda t} - 1\right) + 2Z^*\left(2\exp^{\lambda t} - a(t)\right) + e_1(0)L\left(2 - a(t)\right) \geq 0 \tag{5.158}
$$

By using the definition of $e_1(t)$ in (5.127) the above condition can be rewritten as

$$
b^2 Z^*\left(\exp^{\lambda t} - 1\right) + 4Z^*\left(\exp^{\lambda t}\right) - \frac{2D(0)}{C(0)}\left(2 - a(t)\right) \geq 0 \tag{5.159}
$$

which is always true since $D(0) < 0$, $C(0) > 0$, $Z^* > 0$, $\exp^{\lambda t} \geq 1$ and $a(t) \in [2, \infty)$ (as shown in Appendix D.1).

Concerning $Z(t)$, its derivative can change of sign, so its monotonicity is not ensured. Indeed, $Z(t)$ will be monotonic under the same conditions that $e_1(t)$ is monotonic too. When $e_1(t)$ is not monotonic, a unique peak will appear also in $Z(t)$.

The final coordinates of $\mathbf{X}$ are obtained by calculating the limit when time approaches infinity

$$
\lim_{t\to\infty} X(t) = e_3(0)G \tag{5.160}
$$

$$
\lim_{t\to\infty} Y(t) = e_2(0)G \tag{5.161}
$$

$$
\lim_{t\to\infty} Z(t) = \frac{e_1(0)L - 2Z^* + h(0)Z^*}{h(0)} \tag{5.162}
$$

with

$$G = \frac{1}{2b^3 h(0)} \left(-2b^2 Z^* \left(h(0) \arctan\left(\frac{2}{b}\right) + b \right) + \pi b^2 Z^* h(0) + 2be_1(0)L\left(2 - h(0)\right) \right)$$

(5.163)

Note that by using the definition of e_1 in function of the object parameters in (5.127) and that $h = 2/C$ we have

$$\lim_{t \to \infty} Z(t) = D(0) + Z^*$$

(5.164)

where $D(0)$ is the initial distance between the object plane and the camera origin.

In summary, we can state that a complete analytic model describing the behaviour of the system when using the constant control law has been obtained.

5.7.3 Local asymptotic stability analysis under laser-cross misalignment

The closed-loop equation of the system in presence of calibration errors becomes strongly coupled so that it is not possible to develop the global asymptotic stability analysis under these conditions. We present instead the local asymptotic stability analysis when the laser-cross is not aligned with the camera frame.

Misalignment consisting of a translation

By using the model parameters in Table B.1 we can calculate the interaction matrices of the inverses of y_1, x_2, y_3 and x_4 for the desired state taking into account that the laser-cross is displaced $^C\mathbf{T}_L = (t_x, t_y, t_z)$ from the camera origin. Then, the interaction matrix of $\mathbf{s}$ can be computed obtaining

$$\mathbf{L_s}(e^*) = \begin{pmatrix} 0 & 0 & \frac{2L}{t_y^2 - L^2} & 0 & -\frac{2Lt_x}{t_y^2 - L^2} & 0 \\ 0 & 0 & -\frac{2t_y}{t_y^2 - L^2} & -2 & \frac{2t_x t_y}{t_y^2 - L^2} & 0 \\ 0 & 0 & -\frac{2t_x}{t_x^2 - L^2} & -\frac{2t_x t_y}{t_x^2 - L^2} & 2 & 0 \end{pmatrix}$$

(5.165)

The local asymptotic stability analysis of the system under this type of misalignment consists in studying the product of matrices $\mathbf{M}(e^*) = \mathbf{L_s}(e^*)\widehat{\mathbf{L_s}}^+$ which is

$$\mathbf{M}(e^*) = \begin{pmatrix} \frac{L^2}{L^2 - t_y^2} & 0 & \frac{t_x L}{L^2 - t_y^2} \\ -\frac{t_y L}{L^2 - t_y^2} & 1 & -\frac{t_x t_y}{L^2 - t_y^2} \\ -\frac{t_x L}{L^2 - t_x^2} & -\frac{t_x t_y}{L^2 - t_x^2} & 1 \end{pmatrix}$$

(5.166)

The eigenvalues of $\mathbf{M}(\mathbf{e}^*)$ are

$$
\sigma_1 = \frac{L^2}{L^2 - t_y^2}
$$

$$
\sigma_2 = \frac{L^2 - t_x^2 + \sqrt{t_x^2(t_x^2 - L^2)}}{L^2 - t_x^2}
$$

$$
\sigma_3 = \frac{L^2 - t_x^2 - \sqrt{t_x^2(t_x^2 - L^2)}}{L^2 - t_x^2}
$$

so that imposing the positiveness of σ_1 we have that $L^2 - t_y^2 > 0$ which means that

$$
|t_y| < L \tag{5.167}
$$

When imposing the positiveness of σ_2 and σ_3 we must deal with two hypothesis, one assuming $L^2 - t_x^2 > 0$ and the other $L^2 - t_x^2 < 0$. Let us develop both hypothesis:

- *hypothesis 1:* $L^2 - t_x^2 > 0$. Imposing the positiveness of σ_2 and σ_3 according to this assumption leads to

$$
L^2 - t_x^2 > 0 \Rightarrow \mathrm{Re}(\sigma_2) = \mathrm{Re}(\sigma_3) = 1 \tag{5.168}
$$

  since

$$
\sqrt{t_x^2(t_x^2 - L^2)} = \sqrt{t_x^2(-t_x^2 + L^2)}\sqrt{i} \tag{5.169}
$$

- *hypothesis 2:* $L^2 - t_x^2 < 0$. In this case, imposing the positiveness according to the second hypothesis we obtain

$$
L^2 - t_x^2 < 0 \Rightarrow \sigma_2 > 0 \Leftrightarrow L^2 - t_x^2 + \sqrt{t_x^2(t_x^2 - L^2)} < 0 \tag{5.170}
$$

  which is never true as can be seen by developing the condition as follows

$$
\begin{aligned}
\sqrt{t_x^2(t_x^2 - L^2)} &< -L^2 + t_x^2 \\
t_x^2(t_x^2 - L^2) &< (t_x^2 - L^2)^2 \\
t_x^2 &< t_x^2 - L^2 \\
0 &< -L^2
\end{aligned} \tag{5.171}
$$

Therefore, the right hypothesis is

$$
L^2 - t_x^2 > 0 \tag{5.172}
$$

which imposes that

$$
|t_x| < L \tag{5.173}
$$

Note that the stability domain when using these visual features is a little bit more restricted than when using directly image point coordinates and therefore, than when using the normalised area and angles approach. Concretely, the misalignment of the centre of the laser-cross projected onto the camera plane $Z = 0$ must be included in the square circumscribed by the circle $t_x^2 + t_y^2 < 2L^2$, which was the error tolerated when using image points (see Figure 5.16).

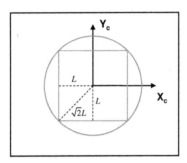

Figure 5.16: Local asymptotic stability areas for the projection of the laser-cross centre onto the plane $Z = 0$.

Misalignment consisting of individual rotations

First of all, let us consider a single rotation ψ of the laser-cross around the X axis of the camera frame. In this case, matrix $\mathbf{L_s}(\mathbf{e}^*)$ is calculated from the model parameters in Table B.2. Then, the eigenvalues of $\mathbf{M} = \mathbf{L_s}(\mathbf{e}^*)\widehat{\mathbf{L_s}}^+$ are

$$
\begin{aligned}
\sigma_1 &= 1 \\
\sigma_2 &= f(L, Z^*, \psi) \\
\sigma_3 &= f(L, Z^*, \psi)
\end{aligned}
$$

the explicit expressions of σ_1 and σ_2 are too complex to be included here. In Figure 5.17a the distribution of σ_2 in function of the rotation ψ and the depth to the object Z^* for $L = 0.15$ m is plotted. In Figure 5.17b a particular case of σ_2 for $Z^* = 1.1$ m. As can be seen, the positiveness of σ_2 is ensured for almost all angle values. The same plots are shown in Figure 5.18 for σ_3. As can be seen, the positiveness of this eigenvalue is not always ensured and depends on the rotation angle.

In the case of a rotation θ of the laser-cross around the Y axis of the camera, also two complex eigenvalues appear, whose distributions are plotted in Figure 5.19 and Figure 5.20, showing that the rotation around the Y axis is better tolerated by the system.

Finally, if a rotation ϕ around the Z axis of the camera is applied to the laser-cross,

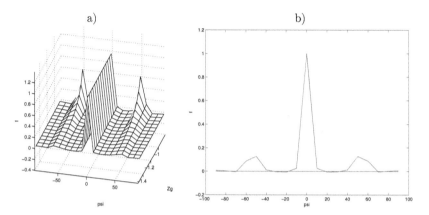

Figure 5.17: a) Rotation around X axis σ_2 in function of ψ (degrees) and $Zg = Z^*$ b) σ_2 in function of ψ (degrees) for the case $Zg = 1.1$ m

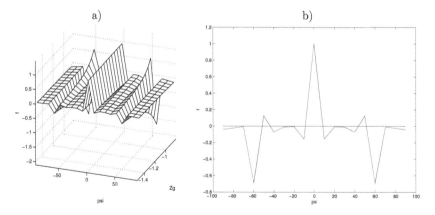

Figure 5.18: Rotation around X axis a) σ_3 in function of ψ (degrees) and $Zg = Z^*$ b) σ_3 in function of ψ (degrees) for the case $Zg = 1.1$ m

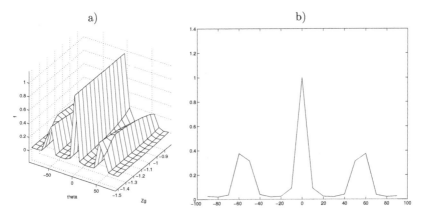

Figure 5.19: Rotation around Y axis a) σ_2 in function of θ (degrees) and $Zg = Z^*$ b) σ_2 in function of θ (degrees) for the case $Zg = 1.1$ m

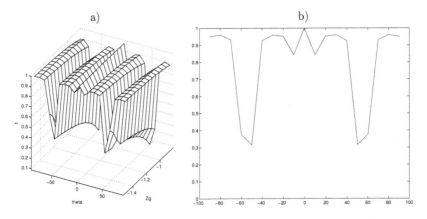

Figure 5.20: Rotation around Y axis a) σ_3 in function of θ (degrees) and $Zg = Z^*$ b) σ_3 in function of θ (degrees) for the case $Zg = 1.1$ m

the eigenvalues obtained from $\mathbf{M}$ are

$$\sigma_1 = \frac{1}{\cos\phi}$$
$$\sigma_2 = 1 + i\tan\phi$$
$$\sigma_2 = 1 - i\tan\phi$$

$$(5.174)$$

imposing the positiveness of the first one, we have that the rotation must be included in $\phi \in [-\pi/2, \pi/2]$. Note that the other eigenvalues are complex numbers and that their real part is always positive. Therefore, rotation of the laser-cross around the optical axis of the camera does not affect the local asymptotic stability of the system.

In summary, the approach based on this non-linear combination of image points is less robust against individual rotations of the laser-cross than the image points based approach and the normalised area and angles approach. In the following section we show how to overcome this problem by improving the set of visual features.

5.7.4 Making features robust against laser-cross misalignment

In this section we present a simple method to enlarge the robustness domain of the features against laser-cross misalignment. The goal is to define a corrected set of visual features $\mathbf{s}'$ which is analytically and experimentally robust against laser-cross misalignment. Figure 5.21 shows the image point distribution in the desired state when different types of misalignment take place (the 4 lasers have the same relative orientation). As can be seen, a general misalignment of the laser-cross produces that the polygon enclosing the 4 points in the desired image appears rotated and translated from the image centre as shown in Figure 5.22a.

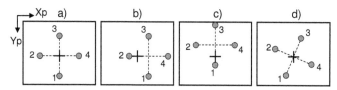

Figure 5.21: Effects of laser-cross misalignment in the desired image. a) Ideal image. b) The laser-cross is horizontally displaced or rotated around Y_C. c) The laser-cross is vertically displaced or rotated around Y_C. d) Laser-cross rotated around Z_C.

In fact, the set of visual features $\mathbf{s} = (a_n, \alpha_{13n}, \alpha_{24n})$ is robust against laser-cross misalignment since both the area and the angles are invariant to the location and orientation of the polygon enclosing the 4 points in the image. Therefore, the corrected set of visual features $\mathbf{s}'$ must be also unaffected by this type of planar transformation. The idea consists in defining an image point transformation composed of a planar transformation and a translation which minimises the misalignment observed in the image. This image transformation will be constrained as follows: in absence of laser-cross misalignment, the corrected set of visual features $\mathbf{s}'$ must be equal to the uncorrected one $\mathbf{s}$. Hence, in the ideal case the results concerning the global asymptotic stability and camera trajectory concerning $\mathbf{s}$ will also hold for $\mathbf{s}'$.

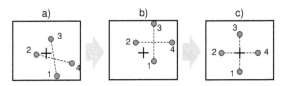

Figure 5.22: Image points correction. a) Desired image under a general misalignment of the laser-cross. b) Image points after applying the transformation $\mathbf{T}$. c) Image points after transformation $\mathbf{T}$ and translation $-\mathbf{x}_g$.

First of all, we eliminate the misalignment exhibited by the polygon in Figure 5.21d which is produced when the laser-cross is rotated around the optical axis. Let us define the following unitary vectors

$$\mathbf{x}_{42} = \begin{pmatrix} x_{42} \\ y_{42} \end{pmatrix} = \frac{\mathbf{x}_4^* - \mathbf{x}_2^*}{\|\mathbf{x}_4^* - \mathbf{x}_2^*\|}, \quad \mathbf{x}_{13} = \begin{pmatrix} x_{13} \\ y_{13} \end{pmatrix} = \frac{\mathbf{x}_1^* - \mathbf{x}_3^*}{\|\mathbf{x}_1^* - \mathbf{x}_3^*\|} \tag{5.175}$$

Then, a simple $2D$ transformation matrix of the form

$$\mathbf{T} = \begin{bmatrix} \mathbf{x}_{24} & \mathbf{x}_{13} \end{bmatrix}^{-1} = \frac{1}{x_{42}y_{13} - x_{13}y_{42}} \begin{pmatrix} y_{13} & -x_{13} \\ -y_{42} & x_{42} \end{pmatrix} \tag{5.176}$$

is defined so that $\mathbf{T}$ uses the desired image points in order to align the unitary vector corresponding to $\mathbf{x}_4 - \mathbf{x}_2$ with the image axis X_p and the unitary vector corresponding to $\mathbf{x}_1 - \mathbf{x}_3$ with the image axis Y_p. Let us note the transformed image points as follows

$$\mathbf{x}_i'' = \mathbf{T}\mathbf{x}_i \tag{5.177}$$

The result of applying the transformation matrix $\mathbf{T}$ to the misaligned image points of Figure 5.22a is shown in Figure 5.22b. Then, it only rests to define a translation vector which is able to centre the polygon in the image (see Figure 5.22c). The most intuitive choice is the centre of gravity of the polygon $\mathbf{x}_g$. However, the choice of a suitable expression for $\mathbf{x}_g$ is not trivial as it could be supposed. First of all, $\mathbf{x}_g$ must be computed from the current image. Secondly, we remember that in absence of laser-cross misalignment $\mathbf{s}'$ must be equal to $\mathbf{s}$ so that $\mathbf{x}_i' = \mathbf{x}_i'' = \mathbf{x}_i$. Hence, under ideal conditions $\mathbf{T}$ must be the identity, which is true according to (5.176), and $\mathbf{x}_g$ must be $\mathbf{0}$. Hence, we could intuitively set $\mathbf{x}_g = (1/4)(\mathbf{x}_1 + \mathbf{x}_2 + \mathbf{x}_3 + \mathbf{x}_4)$. However, according to the ideal model parameters (see Table 5.1), the general expression of the laser image points in function of the object parameters are

$$
\begin{aligned}
x_1 &= 0 & y_1 &= -\frac{LC}{BL+D} \\
x_2 &= -\frac{LC}{AL-D} & y_2 &= 0 \\
x_3 &= 0 & y_3 &= -\frac{LC}{BL-D} \\
x_4 &= -\frac{LC}{AL+D} & y_4 &= 0
\end{aligned}
\tag{5.178}
$$

Therefore, we have that

$$
\frac{1}{4}(\mathbf{x}_1'' + \mathbf{x}_2'' + \mathbf{x}_3'' + \mathbf{x}_4'') = -\frac{L^2 AC}{2\left(A^2 L^2 - D^2\right)}
\tag{5.179}
$$

which is only zero when the camera is parallel to the object. Instead of this, we propose to use

$$
\mathbf{x}_g = \frac{1}{2}\left(\begin{array}{c} x_1'' + x_3'' \\ y_2'' + y_4'' \end{array}\right)
\tag{5.180}
$$

Note that this expression is also a measure of the polygon centre of gravity according to Figure 5.22b. Moreover, in the ideal case $\mathbf{x}_g$ is actually $\mathbf{0}$ for any object pose. Then, the corrected image points are obtained as follows

$$
\mathbf{x}_i' = \mathbf{T}\mathbf{x}_i - \mathbf{x}_g
\tag{5.181}
$$

The corrected set of visual features $\mathbf{s}'$ is therefore

$$
\mathbf{s}' = \left(\begin{array}{ccc} y_1'^{-1} - y_3'^{-1} & y_1'^{-1} + y_3'^{-1} & x_2'^{-1} + x_4'^{-1} \end{array}\right)
\tag{5.182}
$$

The global asymptotic stability of the ideal model is also ensured when using $\mathbf{s}'$. In the following sections the robustness of $\mathbf{s}'$ with respect to laser-cross misalignment is proved analytically. Furthermore, the corrected visual features avoid a potential problem of the uncorrected set $\mathbf{s}$. Certainly, since the definition of $\mathbf{s}$ involves the computation of $1/y_1$, $1/x_2$, $1/y_3$ and $1/x_4$, a division by 0 may be reached due to the laser-cross misalignment. Note that this problem does not longer appear in $\mathbf{s}'$ since the corrected image points are symmetrically distributed around the image centre.

5.7.5 Robust visual features: local asymptotic stability analysis under laser-cross misalignment

The study of the global asymptotic stability of the system when using the corrected set of visual features $\mathbf{s}'$ and the constant control law is again too complex. As in the previous approaches, we instead analyse the local asymptotic stability in front of different types of laser-cross misalignment. Hence, we intend to prove the robustness of the new set of visual features in front of such calibration errors.

Misalignment consisting of a translation

Let us first analyse the case when the laser-cross is aligned with the camera frame, but it is displaced from the camera origin according to $^C\mathbf{T}_L = (t_x, t_y, t_z)$. The real interaction matrix for this laser-cross pose evaluated in the desired state $\mathbf{L}_{\mathbf{s}'}(\mathbf{e}^*)$ must be calculated. First we evaluate the interaction matrices in the desired state of the point coordinates y_1, x_2, y_3 and x_4 using the model parameters in Table B.1 evaluated according to the desired state $A = B = 0$, $C = 1$ and $D = -Z^*$. These parameters are also used in order to calculate the 2D transformation defined in (5.176) and in (5.180) in the desired state. The expressions obtained for $\mathbf{T}$ and $\mathbf{x}_g$ are

$$\mathbf{T} = \begin{pmatrix} 1 & 0 \\ 0 & 1 \end{pmatrix}, \quad x_g = \frac{t_x}{Z^*}, \quad y_g = \frac{t_y}{Z^*} \tag{5.183}$$

The interaction matrix of the corrected set of visual features in the desired state is then

$$\mathbf{L}_{\mathbf{s}'}(\mathbf{e}^*) = \begin{pmatrix} 0 & 0 & -\dfrac{2}{L} & -\dfrac{4t_y}{L} & \dfrac{2t_x}{L} & 0 \\ 0 & 0 & 0 & -2 & 0 & 0 \\ 0 & 0 & 0 & 0 & 2 & 0 \end{pmatrix} \tag{5.184}$$

and the product of matrices in the linearised closed-loop equation of the system $\mathbf{M} = \mathbf{L}_{\mathbf{s}'}(\mathbf{e}^*)\widehat{\mathbf{L}_{\mathbf{s}'}}^+$ is

$$\mathbf{M} = \begin{pmatrix} 1 & 2\dfrac{t_y}{L} & \dfrac{t_x}{L} \\ 0 & 1 & 0 \\ 0 & 0 & 1 \end{pmatrix} \tag{5.185}$$

whose eigenvalues are the elements on the main diagonal which are all equal to 1. Therefore, the local asymptotic stability of the system in front of a displacement of the laser-cross is always ensured when using $\mathbf{s}'$ and the constant control law.

Misalignment consisting of individual rotations

We now present the local asymptotic stability analysis when the laser-cross is centred in the camera origin but it is rotated around one of the camera axis. Let us first consider a rotation $\psi \in (-\pi/2, \pi/2)$ around the X axis. The 2D transformation based on $\mathbf{T}$ and $\mathbf{x}_g$ can be obtained from the model parameters in Table B.2.

$$\mathbf{T} = \begin{pmatrix} 1 & 0 \\ 0 & 1 \end{pmatrix}, \quad x_g = 0, \quad y_g = \frac{\sin\psi}{\cos\psi} \tag{5.186}$$

and the interaction matrix in the desired state taking into account this laser-cross pose is

$$\mathbf{L_{s'}}(\mathbf{e}^*) = \begin{pmatrix} 0 & 0 & -\dfrac{2\cos\psi}{L} & -\dfrac{2Z^*\cos\psi\sin\psi}{L\cos\psi} & 0 & 0 \\ 0 & 0 & 0 & -\dfrac{2\cos\psi}{\cos\psi} & 0 & 0 \\ 0 & 0 & 0 & 0 & 2 & 0 \end{pmatrix} \tag{5.187}$$

so that the product of matrices in the closed-loop equation of the system is

$$\mathbf{M} = \begin{pmatrix} \cos\psi & \dfrac{Z^*\cos\psi\sin\psi}{L\cos\psi} & 0 \\ 0 & 1 & 0 \\ 0 & 0 & 1 \end{pmatrix} \tag{5.188}$$

Note that all the eigenvalues (in this case the elements of the main diagonal) are positive since $\psi \in (-\pi/2, \pi/2)$.

In case that the laser-cross is rotated an angle $\theta \in (-\pi/2, \pi/2)$ around the Y camera axis (model parameters in Table B.3), the $2D$ transformation in the desired state is

$$\mathbf{T} = \begin{pmatrix} 1 & 0 \\ 0 & 1 \end{pmatrix}, \quad x_g = -\frac{\sin\theta}{\cos\theta}, \quad y_g = 0 \tag{5.189}$$

while the interaction matrix in the desired state taking into account this laser-cross pose is

$$\mathbf{L_{s'}}(\mathbf{e}^*) = \begin{pmatrix} 0 & 0 & -\dfrac{2}{L} & 0 & -\dfrac{2Z^*\sin\theta}{L\cos\theta} & 0 \\ 0 & 0 & 0 & -2 & 0 & 0 \\ 0 & 0 & 0 & 0 & \dfrac{2\cos\theta}{\cos\theta} & 0 \end{pmatrix} \tag{5.190}$$

so that the product of matrices in the closed-loop equation of the system is

$$\mathbf{M} = \begin{pmatrix} 1 & 0 & -\dfrac{Z^*\sin\theta}{L\cos\theta} \\ 0 & 1 & 0 \\ 0 & 0 & 1 \end{pmatrix} \tag{5.191}$$

Note that all the eigenvalues are also positive.

Finally, let us study the case when the laser-cross is rotated an angle ϕ around the optical axis of the camera (model parameters in Table B.4. In this case we have that

$$\mathbf{T} = \begin{pmatrix} \cos\phi & \sin\phi \\ -\sin\phi & \cos\phi \end{pmatrix}, \quad x_g = 0, \quad y_g = 0 \tag{5.192}$$

$$\mathbf{L_{s'}(e^*)} = \begin{pmatrix} 0 & 0 & -\dfrac{2}{L} & 0 & 0 & 0 \\ 0 & 0 & 0 & -2\cos\phi & 2\sin\phi & 0 \\ 0 & 0 & 0 & 2\sin\phi & 2\cos\phi & 0 \end{pmatrix} \qquad (5.193)$$

while the product of matrices in the closed-loop equation of the system is

$$\mathbf{M} = \begin{pmatrix} 1 & 0 & 0 \\ 0 & \cos\phi & \sin\phi \\ 0 & -\sin\phi & \cos\phi \end{pmatrix} \qquad (5.194)$$

The eigenvalues are

$$\begin{aligned} \sigma_1 &= 1 \\ \sigma_2 &= \cos\phi + \sqrt{\cos^2\phi - 1} \\ \sigma_3 &= \cos\phi - \sqrt{\cos^2\phi - 1} \end{aligned}$$

Note that the real part of σ_2 and σ_3 is $\cos\phi$ so that in order to ensure its positiveness it is only necessary that $\phi \in [-\pi/2, \pi/2]$.

Therefore, the system based on the corrected set of visual features $\mathbf{s}'$ is locally asymptotically stable if the laser-cross is not aligned with the camera frame.

5.7.6   Robust visual features: local asymptotic stability analysis in presence of camera calibration errors

We now present the local asymptotic stability analysis in presence of calibration errors in the intrinsic parameters of the camera. as explained in Section 5.5.3, it is necessary to study the closed-loop equation of the measured task function $\widetilde{\mathbf{e}}$ that in this case is

$$\dot{\widetilde{\mathbf{e}}}^* = -\lambda \mathbf{L_{\widetilde{s}}}(\widetilde{\mathbf{e}}^*)\widehat{\mathbf{L_s}}^+ \widetilde{\mathbf{e}}^* \qquad (5.195)$$

The interaction matrix in the desired state $\widetilde{\mathbf{e}}^*$ is

$$\mathbf{L_{\widetilde{s}}}(\widetilde{\mathbf{e}}^*) = \begin{pmatrix} 0 & 0 & -\dfrac{2K_v}{L} & 0 & 0 & 0 \\ 0 & 0 & 0 & -2K_v & 0 & 0 \\ 0 & 0 & 0 & 0 & 2K_u & 0 \end{pmatrix} \qquad (5.196)$$

so that the product of matrices $\mathbf{M}(\widetilde{\mathbf{e}}^*) = \mathbf{L_s}(\widetilde{\mathbf{e}}^*)\widehat{\mathbf{L_s}}^+$ is

$$\mathbf{M}(\widetilde{\mathbf{e}}^*) = \begin{pmatrix} K_v & 0 & 0 \\ 0 & K_v & 0 \\ 0 & 0 & K_u \end{pmatrix} \qquad (5.197)$$

Note that the eigenvalues are the elements of the main diagonal and are positive if $K_u > 0$ and $K_v > 0$, which is true if and only if $\widetilde{f} > 0$, $\widetilde{k}_u > 0$ and $\widetilde{k}_v > 0$. Similarly to the previous image-based approaches, the system based on $\mathbf{s}'$ is also robust against camera calibration errors if the elements of the main diagonal of $\widetilde{\mathbf{A}}$ are positive.

5.7.7   Simulation results

We now present the simulation results obtained by the corrected set of visual features $\mathbf{s}'$ according to the conditions described in Section 5.4.1.

Ideal system

As has been shown, under ideal conditions, the interaction matrix of $\mathbf{s}'$ can be evaluated from the task function. The results obtained with the non-constant control law based on a perfect estimation of $\widehat{\mathbf{L}_{\mathbf{s}}}$ are plotted in Figure 5.23. Both an exponential decrease of the task function and a monotonic behaviour of the camera velocities are observed. Furthermore, the camera trajectory is almost a straight line in the space. Note that, as expected, the results coincide with the ones obtained by the position-based approach presented in Section 5.4.1. Unlike the position-based approach, this image-based approach does not require the minimisation of the non-linear equations.

The behaviour of the system when using $\mathbf{s}'$ and the constant control law based on $\mathbf{L}_{\mathbf{s}}^*$ is shown in Figure 5.24. Note that both the task function components and the camera velocities are strictly monotonic as expected from the analytic results. We remark that these results also coincide with the ideal behaviour of the position-based approach based on the constant control law presented in Section 5.4.1. In addition to this, the results of $\mathbf{s}'$ are also very similar to the ones given by the image-based approach using $\mathbf{s} = (a_n, \alpha_{13n}, \alpha_{24n})$. This result was already expected from the Taylor approximations shown in (5.126).

Remember that the behaviour of the system when using the constant control law under ideal conditions can also be obtained from the analytic expressions of the task function $\mathbf{e}(t)$, the camera velocities $V_z(t)$, $\Omega_x(t)$, $\Omega_y(t)$ and the trajectory $\mathbf{X}(t)$ developed in Appendix D.1 and in Section 5.7.2. Given the initial object pose parameters $\underline{\mathbf{n}}$ and D and the desired state defined by $\underline{\mathbf{n}} = (0, 0, 1)$, $D = -Z^*$, the task function components at $t = 0$ can be evaluated by using Equation (D.16), Equation (D.11) and Equation (D.12). The initial conditions found are the following

$$
\begin{aligned}
e_1(0) &= 8.5953 \\
e_2(0) &= 0.5359 \\
e_3(0) &= 1.1547
\end{aligned}
$$

The functions $e_1(t)$, $e_2(t)$ and $e_3(t)$ and the corresponding camera velocities $V_z(t)$, $\Omega_x(t)$, $\Omega_y(t)$ in (5.153) have been evaluated in the interval $t \in [0, 30]$ s and are plotted in Figure 5.25a-b. The coordinates of the initial position expressed in the camera frame obtained analytically in (5.156) have been also evaluated in the same interval. The resulting curves

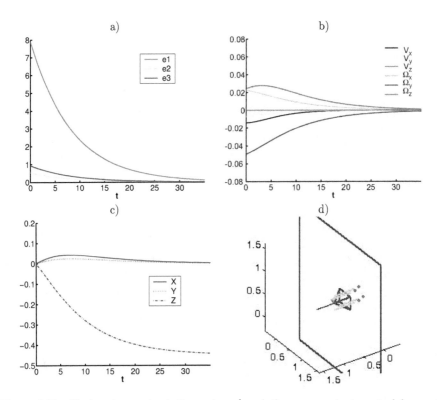

Figure 5.23: Ideal system: simulation using $\mathbf{s}'$ and the non-constant control law.   a) $\mathbf{e} = \mathbf{s} - \mathbf{s}^*$ vs. time (in s). b) Camera velocities (ms/s and rad/s) vs. time. c) Fixed point coordinates in the camera frame. d) Scheme of the camera trajectory.

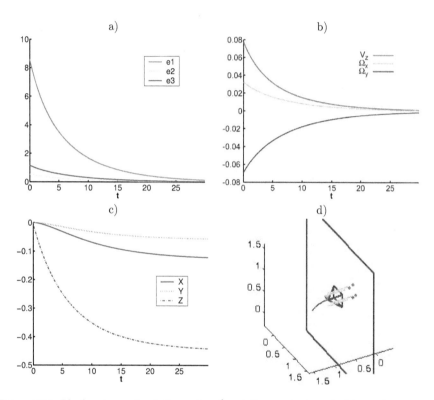

Figure 5.24: Ideal system: simulation using $\mathbf{s}'$ and the constant control law. a) $\mathbf{e} = \mathbf{s} - \mathbf{s}^*$ vs. time (in s). b) Camera velocities (ms/s and rad/s) vs. time. c) Fixed point coordinates in the camera frame. d) Scheme of the camera trajectory.

are plotted in Figure 5.25c and the trajectory of this fixed point in the camera frame is shown in Figure 5.25d. Note that the task function decrease, the camera velocities and the trajectory predicted by the analytic model coincide with the simulation results in Figure 5.24.

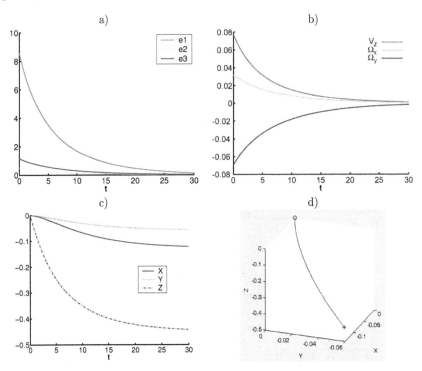

Figure 5.25: Ideal system: analytic behaviour when using $\mathbf{s}'$ and the constant control law. a) $e_1(t)$, $e_2(t)$ and $e_3(t)$ evaluated at $t \in [0, 30]$ s. b) Camera velocities $V_z(t)$ (m/s), $\Omega_x(t)$ and $\Omega_y(t)$ (rad/s) evaluated at $t \in [0, 30]$ s. c) Coordinates of a fixed point (initial position) in the camera frame (in m). d) $3D$ plot of the same point ('o' and '*' are the initial and the final point respectively.

System including laser-cross misalignment and image noise

The behaviour of the system when using $\mathbf{s}'$ and the non-constant control law in presence of large calibration errors and image noise is shown in Figure 5.26. Note that the system is almost unaffected by the calibration errors. We find again that the results are nearly the same that the ones obtained by the position-based approach.

When using the constant control law based on $\mathbf{L_s^*}$, the system is also robust against the laser-cross misalignment, as expected from the local asymptotic stability analysis in

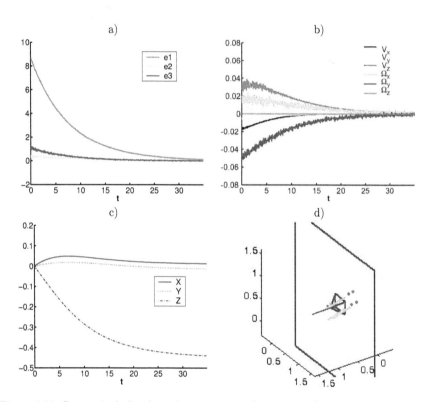

Figure 5.26: System including large laser-cross misalignment and image noise: simulation using $\mathbf{s}'$ and the non-constant control law. a) $\mathbf{e} = \mathbf{s} - \mathbf{s}^*$ vs. time (in s). b) Camera velocities (ms/s and rad/s) vs. time. c) Fixed point coordinates in the camera frame. d) Scheme of the camera trajectory.

presence of such errors. The results are plotted in Figure 5.27. Note that under laser-cross misalignment the simulation results of this approach are still pretty similar to the ones obtained by $\mathbf{s} = (a_n, \alpha_{13n}, \alpha_{24n})$.

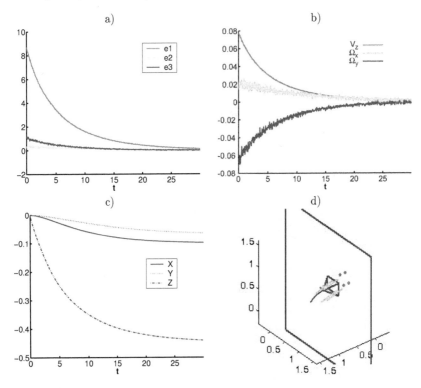

Figure 5.27: System including large laser-cross misalignment and image noise: simulation using $\mathbf{s}'$ and the constant control law. a) $\mathbf{e} = \mathbf{s} - \mathbf{s}^*$ vs. time (in s). b) Camera velocities (ms/s and rad/s) vs. time. c) Fixed point coordinates in the camera frame. d) Scheme of the camera trajectory.

5.8   Experimental results with planar objects

In order to validate the theoretical results and to confirm the simulation results of the different approaches presented in this chapter, real experiments have been carried out. The experimental setup consists of a six-dofs robot manipulator with a camera with focal length 8.5 mm coupled to its end-effector. The images are digitised at 782×582 pixels and the pixel dimensions are about $8.3\mu\text{m} \times 8.3\mu\text{m}$. The laser-cross has been built so that $L = 15$ cm. Such a parameter has been chosen taking into account the robot structure so that the laser-cross can be approximately positioned according to the ideal model, i.e. aligned with the camera frame.

The visual features corresponding to the desired state are calculated through the following learning stage. The camera is positioned with respect to the planar target, already presented in Figure 4.9, by using classic 2D visual servoing. Once the desired position is reached, the lasers are turned on obtaining the desired image point distribution from which the desired visual features $\mathbf{s}^*$ are calculated. This target plane is only used once for obtaining the desired point distribution. Afterwards, the experiments are made with another planar surface containing no visual marks.

The aim of the experiments is to test the behaviour of the control loop when both the laser-cross is positioned according to the ideal model specifications and when a large misalignment between the camera and the laser-cross takes place. Furthermore, during the real experiments a coarse calibration of the camera intrinsic parameters has been used and the direction of all the lasers is not exactly equal, so that the robustness of the approaches against this kind of modelling errors is also tested.

Laser-cross approximately aligned with the camera

In the first experiment, the laser-cross has been approximately aligned with the camera. The desired depth is $Z^* = 60$ cm, while in the initial position the camera is at a distance to the plane of 105 cm and its orientation is defined by $\alpha_x = -20°$ and $\alpha_y = 20°$. The image corresponding to the initial state and the desired image point distribution is shown in Figure 5.28a. On the other hand, Figure 5.28b shows the trace of the laser points in the image from the initial state to the final one.

Indeed, a perfect alignment of the camera and the laser-cross is not possible since we do not exactly know neither the camera origin location nor the orientation of its axis. This is evident by looking at the initial and desired images. As can be seen, the laser points do not exactly lie onto the image axis and their traces from the initial to the desired position (which shows us the epipolar line of each laser) are not perfectly parallel to the axis. Furthermore, it is neither possible to ensure that all the 4 lasers have the same exact orientation (which causes that the epipolar lines do not intersect in a unique point).

The position-based approach has not been implemented since it is equivalent to the 2D approach based on $\mathbf{s}'$. Furthermore, the latter has the advantage that it is less time consuming since no numerical minimisation of non-linear equations is required.

First of all, Figure 5.29 presents the system response when using image points as visual features. As can be seen, since the laser-cross misalignment is small enough, the system converges showing a nice decrease on the visual feature errors and the norm of the task function (even if it is not a pure exponential decrease, as expected from the expression of $\mathbf{L_s^*}$ which depends on $1/Z^*$). On the other hand, note that the camera velocities generated by the constant control law are not monotonic, specially the rotational ones.

Figure 5.30 shows the results when using the set of visual features based on the normalised area and angles. As expected, both the task function and the camera velocities better fit an exponential decrease. Furthermore, a linear mapping from task function space to camera velocities is almost exhibited.

Very similar results are obtained with the corrected version of the decoupled set of visual features $\mathbf{s}'$, which are presented in Figure 5.31. In this case, the results when using the constant control law are plotted. We can observe the monotonic decrease of the task function and the camera velocities as predicted by the analytic model. No major differences are appreciated with respect to the approach based on normalised area and angles.

In Figure 5.32 the results when using $\mathbf{s}'$ and the non-constant control law are shown. In this case, a pure exponential decrease of the task function is expected. Note however, that the actual behaviour is not monotonic, which implies that this type of control law is a bit more sensitive to the lasers directions and the camera calibration errors. Nevertheless, the system converges with no major problems. Note also the non-monotonic camera velocities generated by the control law. We must also mention that when using the non-constant control law the computation time required at each iteration is higher since the pseudoinverse of the estimated $\widehat{\mathbf{L_s}}$ must be calculated.

Large misalignment between the camera and the laser-cross

The same experiment has been repeated by introducing a large misalignment between the laser-cross and the camera. Concretely, the laser-cross has been displaced from the camera origin about 6 cm in the sense of the $-X$ axis of the camera frame. Furthermore, it has been rotated about 7° around the Z axis (the rotation introduced about the X and Y axis are much smaller). Such a large misalignment is clearly observed in the initial and desired image points distribution shown in Figure 5.33a. The final image is shown in Figure 5.33b.

Under these conditions, only the approaches based on normalised area and angles, and the corrected version of the decoupled set of visual features have succeeded. On the other hand, the image-based approach based on $\mathbf{s} = (y_1, x_2, y_3, x_4)$ has diverged as expected from the simulation results. In Figure 5.34 the results when using the constant control law based on $\mathbf{s} = (a_n, \alpha_{13n}, \alpha_{24n})$ are shown. On the other hand, when using the corrected set of visual features, the original image is transformed at each iteration producing the image trace plotted in Figure 5.33c. Then, Figure 5.35 presents the results when using $\mathbf{s}'$. As can be seen, even with such a large misalignment of the laser-cross, both approaches still obtain almost a monotonic decrease in the task function as well as an almost monotonic decrease in the camera velocities. Therefore, the large convergence domain of these approaches expected from the analytic results is here confirmed.

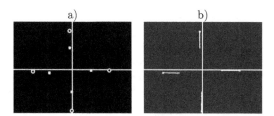

Figure 5.28: Experiment using an approximated alignment. a) Initial image (solid dots) including image axis and the desired position of each laser point (circles). b) Final image with the trace of each laser point from its initial position to its final position.

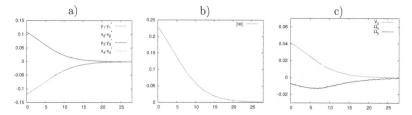

Figure 5.29: Approximated alignment: experiment using $\mathbf{s} = (y_1, x_2, y_3, x_4)$ and the constant control law. a) $\mathbf{s} - \mathbf{s}^*$ vs. time (in s). b) Norm of the task function vs. time. c) Camera velocities (ms/s and rad/s) vs. time.

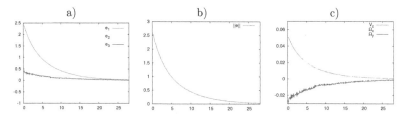

Figure 5.30: Approximated alignment: experiment using $\mathbf{s} = (a_n, \alpha_{13n}, \alpha_{24n})$ and the constant control law. a) $\mathbf{e} = \mathbf{s} - \mathbf{s}^*$ vs. time (in s). b) Norm of the task function vs. time. c) Camera velocities (ms/s and rad/s) vs. time.

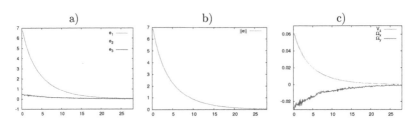

Figure 5.31: Approximated alignment: experiment using $\mathbf{s} = (y_1'^{-1} - y_3'^{-1}, y_1'^{-1} + y_3'^{-1}, x_2'^{-1} + x_4'^{-1})$ and the constant control law. a) $\mathbf{e} = \mathbf{s} - \mathbf{s}^*$ vs. time (in s). b) Norm of the task function vs. time. c) Camera velocities (ms/s and rad/s) vs. time.

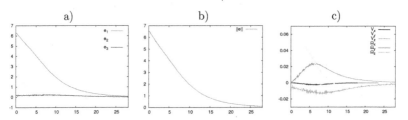

Figure 5.32: Approximated alignment: experiment using $\mathbf{s} = (y_1'^{-1} - y_3'^{-1}, y_1'^{-1} + y_3'^{-1}, x_2'^{-1} + x_4'^{-1})$ and the non-constant control law. a) $\mathbf{e} = \mathbf{s} - \mathbf{s}^*$ vs. time (in s). b) Norm of the task function vs. time. c) Camera velocities (ms/s and rad/s) vs. time.

When using $\mathbf{s}'$ and the non-constant control law, the system has not been able to converge to the desired position since the robot has reached a joint limit. This fact has been also observed when using other initial positions. It seems that the non-linearities in the camera velocities produced by such a control law become stronger due to errors in the lasers directions when the laser-cross is largely misaligned. Hence, some of the demanded robot motions are unfeasible or usually bring the robot very close to some joint limits. Therefore, we confirm that designing decoupled visual features which vary proportionally to the corresponding controlled dof is a good strategy to obtain suitable camera trajectories [Mahony et al., 2002].

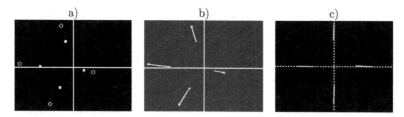

Figure 5.33: Experiment using a large misalignment. Initial image (solid dots) including image axis and the desired position of each laser point (circles). b) Final image with the trace of each laser point from its initial position to its final position. c) Corrected image from the initial to the desired position (the dashed lines show the image axis).

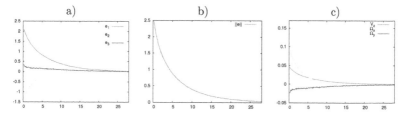

Figure 5.34: Large misalignment: experiment using $\mathbf{s} = (a_n, \alpha_{13n}, \alpha_{24n})$ and the constant control law. a) $\mathbf{e} = \mathbf{s} - \mathbf{s}^*$ vs. time (in s). b) Norm of the task function vs. time. c) Camera velocities (ms/s and rad/s) vs. time.

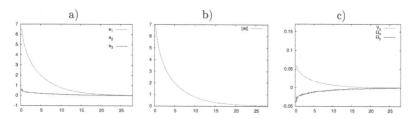

Figure 5.35: Large misalignment: experiment using $\mathbf{s} = ({y_1'}^{-1} - {y_3'}^{-1}, {y_1'}^{-1} + {y_3'}^{-1}, {x_2'}^{-1} + {x_4'}^{-1})$ and the constant control law. a) $\mathbf{e} = \mathbf{s} - \mathbf{s}^*$ vs. time (in s). b) Norm of the task function vs. time. c) Camera velocities (ms/s and rad/s) vs. time.

5.9   Summary

In this section we briefly summarise the different approaches presented along the chapter for plane-to-plane positioning. The analytic results concerning the stability under ideal conditions and under calibration errors are summarised in Table 5.2. The results provided in the table take into account the control based on the interaction matrix evaluated in the desired state $\mathbf{L_s^*}$. When using a non-constant control law, the global asymptotic stability has only been proven when the estimation of $\mathbf{L_s}$ at each iteration is perfect, that is, when the conditions of the ideal model hold and there are no calibration errors.

Position-based approach: the input of the control law are the parameters of the object plane equation γ, β and α. The interaction matrix of such parameters shows a nice decoupling from the rotational to the translational part. Calculating these parameters requires to solve at each iteration a system of non-linear equations. This can be done numerically with a minimisation algorithm which can be sensible to fall into local minima. Therefore, the convergence of the approach cannot be proven. Nevertheless, the camera trajectory obtained is almost a straight line and the task function shows an exponential decrease under ideal conditions.

Image points approach: the inputs of the control law are the image point coordinates (y_1, x_2, y_3, x_4). The stability of the system is not ensured in presence of a displacement of the laser-cross with respect to the camera centre. Furthermore, the camera velocities are non-monotonic.

Normalised area and angles approach: the input of the control law are the geometric image-based features $(a_n, \alpha_{13n}, \alpha_{24n})$, which have shown a strong robustness against calibration errors both analytically (through the local asymptotic stability analysis) and experimentally. Simulations and experiments have shown that both the task function and the camera velocities are monotonic. However, due to the complexity of the features no analytic results concerning the global asymptotic stability have been provided.

Decoupled features approach: the input of the control law are the image-based features

$$\mathbf{s} = \left(y_1^{-1} - y_3^{-1}, y_1^{-1} + y_3^{-1}, x_2^{-1} + x_4^{-1} \right) \qquad (5.198)$$

This set of features decouples the rotational from the translational dof in all the workspace and the interaction matrix can be entirely expressed in terms of the task function approach. This is possible since these features are proportional to the object parameters γ, β and α. It has been possible to demonstrate the global asymptotic stability under ideal conditions when using both a constant control law based on $\mathbf{L_s^*}$ and when using an estimation of $\mathbf{L_s}$ (which does not require to reconstruct the object). When using the non-constant control law a pure exponential decrease of the task function is obtained under ideal conditions. In the case of the constant control law, a monotonic decrease is obtained for the visual features controlling the rotational subsystem, and the behaviour of the feature controlling the depth is either monotonic or it presents a unique peak. The system is however quite sensitive to laser-cross misalignment.

Corrected decoupled features approach: a simple planar transformation applied to the image allows the robustness domain in presence of calibration errors to be enlarged. The corrected set of visual features $\mathbf{s}'$ has nice robustness properties and obtains the same results than the uncorrected version under ideal conditions.

Table 5.2: Stability analysis of the different approaches

Visual Features	Ideal system: stability		Laser-cross misalignment: local asymptotic stability				intrinsic errors: local as. stab.
	Global	Local	(t_x,t_y,t_z)	Rot(X,ψ)	Rot(Y,θ)	Rot(Z,ϕ)	
(γ,β,α)	$\checkmark$	$\checkmark$	?	?	?	?	?
(y_1,x_2,y_3,x_4)	?	$\checkmark$	$t_x^2+t_y^2<2L^2$	$\checkmark$	$\checkmark$	$\checkmark$	$\checkmark$
$(a_n,\alpha_{13n},\alpha_{24n})$	?	$\checkmark$	$\checkmark$	$\checkmark$	$\checkmark$	$\checkmark$	$\checkmark$
$\mathbf{s}$	$\checkmark$	$\checkmark$	$\lvert t_{x,y}\rvert<L$	restricted	$\checkmark$	$\checkmark$	?
$\mathbf{s}'$	$\checkmark$	$\checkmark$	$\checkmark$	$\checkmark$	$\checkmark$	$\checkmark$	$\checkmark$

5.10 Positioning task with respect to non-planar objects

The number of pure image-based approaches able to deal with non-planar objects is quite reduced and they usually have some limitations. In some cases, they are model-based since they are only valid for certain objects. As example we have the approaches presented by Espiau et al. [Espiau *et al.*, 1992] which allow the camera to be positioned with respect to cylinders and spheres. There are other approaches that are considered model-free. For example, a typical example are the approaches based on image points, but they require an estimation of the depth distribution [Benhimane and Malis, 2003; Schramm *et al.*, 2004]. Furthermore, such estimation must be accurate enough for ensuring the convergence [Malis and Rives, 2003]. Another example is the contour approach presented in [Collewet and Chaumette, 2000] which does not require depth information and it is able to deal with non-planar objects. However, the object's curvature must be weak and only binary objects have been taken into account.

The structured light emitter proposed in this chapter has been designed for positioning the robot with respect to planar objects. However, it is interesting to see what happens when the object is actually non-planar. In the following sections, the case of quadric objects is considered. This type of objects have been addressed because some analytic predictions can be formulated. Furthermore, simulations and experiments illustrating this case are shown.

In next section we investigate whether it is possible to cancel the task function of the decoupled image-based approach proposed in this chapter when the object is a quadric.

5.10.1   Non-planar objects: can the task function cancel?

The decoupled image-based approach is based on the following set of visual features

$$\mathbf{s} = (s_1,\ s_2,\ s_3) = \left(y_1^{-1} - y_3^{-1},\ \ y_1^{-1} + y_3^{-1},\ \ x_2^{-1} + x_4^{-1}\right) \tag{5.199}$$

where s_1 controls the depth and s_2 and s_3 the orientation of the camera with respect to a planar object. Taking into account that $\mathbf{x} = \mathbf{Y}/\mathbf{Z}$ and the lasers distribution around the camera provided by Table 5.1, the normalised image points corresponding to the lasers are

$$
\begin{aligned}
y_1 &= L/Z_1 \\
y_3 &= -L/Z_3 \\
x_2 &= -L/Z_2 \\
x_4 &= L/Z_4
\end{aligned}
\tag{5.200}
$$

Let us hereafter consider two cases: when the object is unknown and when a model of the object is available.

Unknown object model

A possible strategy when the object model is unknown is to consider like if it was a planar object. In this case, all the four laser points are supposed to lie at a depth Z^*, which is the depth included in the model of interaction matrix used in the control law. Under this modelling assumption, the desired laser point distribution in the image is

$$
\begin{aligned}
y_1^* &= L/Z^* \\
y_3^* &= -L/Z^* \\
x_2^* &= -L/Z^* \\
x_4^* &= L/Z^*
\end{aligned}
\tag{5.201}
$$

Therefore, the task function components $\mathbf{e} = \mathbf{s} - \mathbf{s}^*$ can be expressed as

$$e_1 = \frac{1}{y_1} - \frac{1}{y_3} - \left(\frac{1}{y_1^*} - \frac{1}{y_3^*}\right) = \frac{1}{L}(Z_1 + Z_3 - 2Z^*) \tag{5.202}$$

$$e_2 = \frac{1}{y_1} + \frac{1}{y_3} - \left(\frac{1}{y_1^*} + \frac{1}{y_3^*}\right) = \frac{1}{L}(Z_1 - Z_3) \tag{5.203}$$

$$e_2 = \frac{1}{x_2} + \frac{1}{x_4} - \left(\frac{1}{x_2^*} + \frac{1}{x_4^*}\right) = \frac{1}{L}(Z_4 - Z_2) \tag{5.204}$$

According to (5.203) and (5.204) the following conditions hold

$$e_2 = 0 \Leftrightarrow Z_1 = Z_3 \tag{5.205}$$
$$e_3 = 0 \Leftrightarrow Z_2 = Z_4 \tag{5.206}$$

which implies that

$$e_1 = 0 \Leftrightarrow Z_1 = Z_3 = Z^* \tag{5.207}$$

In summary, the task function cancels when $Z_1 = Z_3 = Z^*$ and $Z_2 = Z_4$. The question is: can the task function cancel when the object is non-planar and no model of it is available?

We address this problem by taking quadric objects into account. A quadric can be expressed by the following equation

$$\lambda_1{}^o X^2 + \lambda_2{}^o Y^2 + \lambda_3{}^o Z^2 - M = 0 \tag{5.208}$$

being $M > 0$ and noting that $(^o X, {}^o Y, {}^o Z)$ is a vector expressed in the canonic frame of the quadric $\{O\}$. For a comprehensive taxonomy of the different quadrics we refer to [Audin, 2002].

It is easy to demonstrate that, for any quadric object, there is at minimum a position of the camera where the task function cancels. Let the camera frame $\{C\}$ have the same orientation that the canonic frame $\{O\}$ and the origin of $\{O\}$ being expressed in the camera frame as $^C(0, 0, O_z)$. Figure 5.36 illustrates this case when the quadric is an sphere.

In this particular configuration, the quadric is expressed in the camera frame by the equation

$$\lambda_1 X^2 + \lambda_2 Y^2 + \lambda_3 (Z - O_z)^2 - M = 0 \tag{5.209}$$

The (X, Y) coordinates of the laser points projected on the object are provided by the lasers distribution in Table 5.1 being

$$\begin{array}{llll}
X_1 &=& 0 \qquad & Y_1 &=& L \\
X_2 &=& -L \qquad & Y_2 &=& 0 \\
X_3 &=& 0 \qquad & Y_3 &=& -L \\
X_4 &=& L \qquad & Y_4 &=& 0
\end{array} \tag{5.210}$$

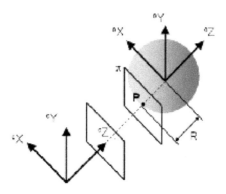

Figure 5.36: Example of relative pose between the camera and a sphere for which the task function cancels.

The depth of the points are obtained from (5.209) and there are two solutions

$$Z = O_z \pm \frac{\sqrt{\lambda_3(M - \lambda_1 X^2 - \lambda_2 Y^2)}}{\lambda_3} \tag{5.211}$$

Considering the first solution

$$Z = O_z + \frac{\sqrt{\lambda_3(M - \lambda_1 X^2 - \lambda_2 Y^2)}}{\lambda_3} \tag{5.212}$$

the depths of the laser points are obtained by using (5.210)

$$\begin{aligned} Z_1 = Z_3 &= O_z + \frac{1}{\lambda_3}\sqrt{\lambda_3(M - \lambda_2 L^2)} \\ Z_2 = Z_4 &= O_z + \frac{1}{\lambda_3}\sqrt{\lambda_3(M - \lambda_1 L^2)} \end{aligned} \tag{5.213}$$

Therefore, in this position, the task function components e_2 and e_3 are always cancelled for any type of quadric object with depth distribution described by (5.212). Then, according to (5.202) the task function component e_1 is

$$e_1 = \frac{1}{L}(Z_1 + Z_3 - 2Z^*) = \frac{2}{L}\left(O_z - Z^* + \frac{1}{\lambda_3}\sqrt{\lambda_3(M - \lambda_2 L^2)}\right) \tag{5.214}$$

which is 0 only when the camera is at the right distance of the quadric, given by

$$O_z = Z^* - \frac{1}{\lambda_3}\sqrt{\lambda_3(M - \lambda_2 L^2)} \tag{5.215}$$

In this case the laser depths are

$$
\begin{aligned}
Z_1 = Z_3 &= Z^* \\
Z_2 = Z_4 &= Z^* + \frac{1}{\lambda_3} \left(\sqrt{\lambda_3(M - \lambda_1 L^2)} - \sqrt{\lambda_3(M - \lambda_2 L^2)} \right)
\end{aligned}
\tag{5.216}
$$

If the second solution in (5.211) is considered the depth distribution of the object is

$$
Z = O_z - \frac{\sqrt{\lambda_3(M - \lambda_1 X^2 - \lambda_2 Y^2)}}{\lambda_3}
\tag{5.217}
$$

and the depths of the laser points are

$$
\begin{aligned}
Z_1 = Z_3 &= O_z - \frac{1}{\lambda_3} \sqrt{\lambda_3(M - \lambda_2 L^2)} \\
Z_2 = Z_4 &= O_z - \frac{1}{\lambda_3} \sqrt{\lambda_3(M - \lambda_1 L^2)}
\end{aligned}
\tag{5.218}
$$

As can be seen, in this case e_2 and e_3 are also 0. The task component e_1 is

$$
e_1 = \frac{1}{L}(Z_1 + Z_3 - 2Z^*) = \frac{2}{L}\left(O_z - Z^* - \frac{1}{\lambda_3}\sqrt{\lambda_3(M - \lambda_2 L^2)} \right)
\tag{5.219}
$$

being 0 when the object frame origin is in front of the camera at a distance

$$
O_z = Z^* + \frac{1}{\lambda_3}\sqrt{\lambda_3(M - \lambda_2 L^2)}
\tag{5.220}
$$

Then, the laser depths are

$$
\begin{aligned}
Z_1 = Z_3 &= Z^* \\
Z_2 = Z_4 &= Z^* + \frac{1}{\lambda_3}\left(\sqrt{\lambda_3(M - \lambda_2 L^2)} - \sqrt{\lambda_3(M - \lambda_1 L^2)} \right)
\end{aligned}
\tag{5.221}
$$

Note that, according to the laser points depth distribution in (5.213) and the symmetric case in (5.218), it is clear that the four laser points are co-planar, like in the case of a planar object, when $\lambda_1 = \lambda_2$. This case appears for objects of revolution around the Z axis like sphere, ellipsoids of revolution, hyperboloids of revolution and cones of revolutions.

Let us calculate the tangent plane of the quadric at the intersection point $\mathbf{P}$ with the optical axis. The normal to this point is defined by

$$
\mathbf{n} = \left(-\frac{\partial Z}{\partial X}\Big|_{\mathbf{P}}, \ -\frac{\partial Z}{\partial Y}\Big|_{\mathbf{P}}, \ 1 \right)
\tag{5.222}
$$

with $\mathbf{P} = (0, 0, Z_P)$. For both depth distributions in (5.212) and in (5.217) the normal is

$$\mathbf{n} = (0,\ 0,\ 1) \tag{5.223}$$

Therefore, in this position, the camera optical axis has the same orientation than the normal to the tangent plane at $\mathbf{P}$. Furthermore, the depth Z_P is

$$Z_P = O_z + \frac{\sqrt{\lambda_3 M}}{\sqrt{\lambda_3}} \tag{5.224}$$

when the depth distribution is given by (5.212) or

$$Z_P = O_z - \frac{\sqrt{\lambda_3 M}}{\sqrt{\lambda_3}} \tag{5.225}$$

when given by (5.217).

In summary, it has been proven that for any quadric, there is at least one position of the camera for which the task function is cancelled even if the model of the object is unknown and it is supposed to be planar. Furthermore, in this position the camera gets parallel to the object tangent plane at the intersection point between the optical axis and the object. Nevertheless, we do not know whether the visual control approach is able to move the camera to such position for any initial state.

Model of the object available

In this case, as the model of the object is available, the true desired visual features can be calculated and some information about the object shape can be included in the interaction matrix.

Let us consider a concrete example: the goal is to get positioned in front of a quadric object like in the example illustrated in Figure 5.36. Then, the true depths of every projected laser point in the desired state are given either by (5.212) or by (5.217). With these depths, the desired visual features can be calculated by taking into account that

$$\begin{aligned}
y_1^* &= L/Z_1^* \\
y_3^* &= -L/Z_3^* \\
x_2^* &= -L/Z_2^* \\
x_4^* &= L/Z_4^*
\end{aligned} \tag{5.226}$$

In order to include some information about the object shape in the interaction matrix let us use the following procedure. We remember that the interaction matrix of every projected point is

$$\mathbf{L_x} = \frac{1}{\Pi_0}\begin{pmatrix} \dfrac{-AX_0}{Z} & \dfrac{-BX_0}{Z} & \dfrac{-CX_0}{Z} & X_0\varepsilon_1 & X_0\varepsilon_2 & X_0\varepsilon_3 \\[2mm] \dfrac{-AY_0}{Z} & \dfrac{-BY_0}{Z} & \dfrac{-CY_0}{Z} & Y_0\varepsilon_1 & Y_0\varepsilon_2 & Y_0\varepsilon_3 \end{pmatrix}$$

$$(5.227)$$

$$\Pi_0 = \mathbf{n}^\top(\mathbf{X}_0 - \mathbf{x}Z)$$
$$(\varepsilon_1, \varepsilon_2, \varepsilon_3) = \underline{\mathbf{n}} \times (x, y, 1) \tag{5.228}$$

which includes the normal $\mathbf{n} = (A, B, C)$ to the object in the projected laser point. In the case of planar objects, it has been considered that the normal to every projected point was equal. If the object is non-planar and its model is known, the actual normal to each laser point can be used. Let

$$\begin{aligned} \mathbf{n}_1 &= (A_1,\ B_1,\ C_1) \\ \mathbf{n}_2 &= (A_2,\ B_2,\ C_2) \\ \mathbf{n}_3 &= (A_3,\ B_3,\ C_3) \\ \mathbf{n}_4 &= (A_4,\ B_4,\ C_4) \end{aligned} \tag{5.229}$$

be the object normals to each projected laser point. These normals can be used for calculating the interaction matrices $\mathbf{L}_{y_1}$, $\mathbf{L}_{x_2}$, $\mathbf{L}_{y_3}$ and $\mathbf{L}_{x_4}$. And from them, the interaction matrix of the decoupled set of visual features

$$\mathbf{s} = (s_1,\ s_2,\ s_3) = \left(y_1^{-1} - y_3^{-1},\ y_1^{-1} + y_3^{-1},\ x_2^{-1} + x_4^{-1}\right) \tag{5.230}$$

is obtained

$$\mathbf{L_s} = \begin{pmatrix} \frac{A_1C_3+A_3C_1}{-LC_1C_3} & \frac{B_1C_3+B_3C_1}{-LC_1C_3} & \frac{-2}{L} & \frac{C_3B_1Z_1+C_1B_3Z_3}{LC_1C_3} & \frac{Z_1A_1C_3+Z_3A_3C_1}{-LC_1C_3} & \frac{A_1C_3-A_3C_1}{C_1C_3} \\[2mm] \frac{A_3C_1-A_1C_3}{LC_1C_3} & \frac{B_3C_1-B_1C_3}{LC_1C_3} & 0 & \frac{C_3B_1Z_1-2LC_1C_3-C_1B_3Z_3}{LC_1C_3} & \frac{Z_1A_1C_3+Z_3A_3C_1}{-LC_1C_3} & \frac{A_1C_3+A_3C_1}{C_1C_3} \\[2mm] \frac{A_2C_4-A_4C_2}{LC_2C_4} & \frac{B_2C_4-B_4C_2}{LC_2C_4} & 0 & \frac{Z_4B_4C_2-Z_2^2B_2C_4}{LC_2C_4} & \frac{2LC_2C_4+C_4A_2Z_2-C_2A_4Z_4}{LC_2C_4} & \frac{B_4C_2+B_2C_4}{-C_2C_4} \end{pmatrix}$$

$$(5.231)$$

From this expression it is possible to calculate the model of interaction matrix used in the control law by using the normals and depths corresponding to the desired state. Note that the interaction matrix is no longer decoupled unless all the normals are equal to $(0,\ 0,\ 1)$ which is the case of a planar object.

The following sections investigates several examples through simulations and some experiments.

5.10.2   Case of a sphere

A sphere of radius R is represented in its canonic frame $\{O\}$ by the well known cartesian equation

$$^{o}X^2 + {}^{o}Y^2 + {}^{o}Z^2 - R^2 = 0 \tag{5.232}$$

In this case, there are infinite camera positions for which the task function is cancelled. Concretely, the task function can be cancelled for any position where the optical axis direction contains the sphere centre which must be at the coordinates $^{C}(0, 0, O_z)$ with

$$O_z = Z^* + \sqrt{R^2 - L^2} \tag{5.233}$$

Several simulations are hereafter presented showing the behaviour of the image-based decoupled approach. The first two simulations assume that the model of the sphere is not known. Therefore, the desired visual features and the control law are calculated assuming a planar object. The third simulation shows the behaviour of the system when the model of the sphere is known.

First simulation: unknown sphere

The first simulation consists in positioning the camera with respect to a sphere of radius $R = 0.4$ m. The camera initial pose can be described with the aid of Figure 5.36. The origin of the camera in the initial state in the sphere frame is given by $^{o}(0, 0, 1.4)$ m. Then, the orientation of the camera with respect of the sphere is defined by the angles $\alpha_x = -8°$ and $\alpha_y = -6°$ which are expressing the orientation of the optical axis with respect to the plane $^{o}Z = 0$.

The model of the sphere is supposed to be unknown and the desired visual features and the control law are calculated assuming a planar object placed at the desired depth. The initial relative pose camera-sphere is represented in Figure 5.37a. Note that only the visible part of the sphere surface is represented. The initial laser point distribution on the image as well as the desired image point distribution are shown in Figure 5.37b. At the end of the simulation, the camera reaches the position shown in Figure 5.37c.

The behaviour of the task function and the camera velocities during the simulation is shown in Figure 5.37d-e. As can be seen, the behaviour is quite similar to the one obtained with a planar object. The main difference is that the rate of convergence of the rotational subsystem controlled by e_2 and e_3 is faster. This is due to the curvature of the object which provokes larger variations in the task function than if a planar object was used. About the camera trajectory, Figure 5.37f shows the coordinates of the initial position in the camera frame along time. As can be seen, the major variation occurs in depth, while lateral displacements are quite reduced.

In order to evaluate the correctness of the final position we present numeric results corresponding to the final state in Table 5.3. As can be seen, the camera actually reaches a position where it is parallel to the tangent plane of the sphere at point **P**, as shown by the very small values of α_x and α_y and the normal to Π. Note that the task function can

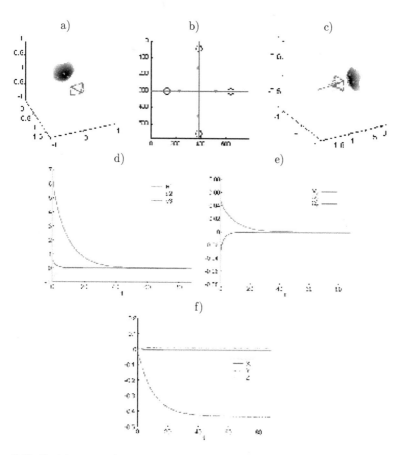

Figure 5.37: Positioning with respect to an unknown sphere (only a portion of the sphere is represented). a) Initial position. b) Initial image point distribution (solid dots) including the laser epipolar lines and the desired image point distribution (circles). c) Camera trajectory till the final position. d) Task function vs. time (in s). e) Camera velocities (in m/s and rad/s). f) Camera initial position (in m) expressed in the camera frame vs. time (in s).

be considered numerically cancelled.

Furthermore, in order to check if the camera has reached the type of position studied in Section 5.10.1, the expected laser depths and the expected depth Z_P^* to point **P** are also shown in Table 5.3. The expected laser depths have been calculated with (5.221). The expected depth Z_P^* has been calculated by using (5.220) and then (5.225). As can be seen in the table, the depths corresponding to the final state of the simulation are equal to the ones predicted analytically.

Table 5.3: Final results when positioning with respect to an unknown sphere.

Normal to Π:	$^c(0, 0, 1)$
(α_x, α_y) with respect to Π:	$(-2.905\mathrm{e} - 5°, -4.444\mathrm{e} - 5°)$
Z_P:	0.5709 m
Z_P^*:	0.5708 m
Task function **e**:	$(0.6823\mathrm{e} - 3, \ -0.0012\mathrm{e} - 3, \ 0.0016\mathrm{e} - 3)$
Lasers depths:	$(0.6001, \ 0.6001, \ 0.6001, \ 0.6001)$ m
Expected laser depths:	$(0.6, \ 0.6, \ 0.6, \ 0.6)$ m

Second simulation: unknown sphere including a fixation point

As said before, in the case of a sphere, there are infinite positions where the task function can be cancelled so that the camera gets parallel to the sphere tangent plane to a certain point **P**. We now investigate whether it is possible to choose such point **P**. We consider that a given point of the sphere can be tracked during the camera motion. Therefore, this point can be used as fixation point and the goal consists in centering point **P** in the image so that the camera gets parallel to the tangent plane in **P**. Again, we suppose that the model of the sphere is unknown. We suggest two ways of achieving such a task:

- Stacking the interaction matrix $\mathbf{L_x}$ of the fixation point in the interaction matrix of the decoupled image-based approach.

- Centering the fixation point by using a secondary task.

The first option has the disadvantage that the decoupling of the plane-to-plane positioning approach is lost. Furthermore, simulations have shown that a better camera trajectory is obtained if a secondary task is used according to the redundancy approach by Espiau et al. [Espiau *et al.*, 1992]. According to the redundancy approach definition, if the real interaction matrix is known at each iteration, only camera motions belonging to its kernel will be allowed for the secondary task. However, when a static estimation of the interaction matrix is only available, as in our case since $\mathbf{L_s^*}$ is used, the secondary task can slightly perturb the primary one. Nevertheless, with this formalism we reduce the influence of the secondary task over the plane-to-plane positioning goal. Note that given the kernel of $\mathbf{L_s^*}$ the camera motions that the secondary task can generate are V_x, V_y and Ω_z. The control law taking into account the secondary task is

$$\mathbf{v} = -\lambda_1 \left(\widehat{\mathbf{L_s}}^+ (\mathbf{s} - \mathbf{s^*}) + \lambda_2 \left(\mathbf{I_6} - \widehat{\mathbf{L_s}}^+ \widehat{\mathbf{L_s}} \right) \widehat{\mathbf{L_x}}^+ (\mathbf{x} - \mathbf{x^*}) \right) \tag{5.234}$$

where $\mathbf{x}$ are the normalised coordinates of the fixation point and $\widehat{\mathbf{L_x}}$ is the interaction matrix in (5.2) evaluated for the desired state, i.e. when the point is centred in the image.

The same conditions described in the previous simulation have been used in this case. Figure 5.38a shows a representation of the initial position of the camera with respect to the sphere and the fixation point drawn in green. The initial and final image point distribution as well as the initial position in the image of the fixation point are represented in Figure 5.38b. As can be seen in Figure 5.38c, the camera trajectory differs from the one obtained without fixation point since the final position is no longer the same. Note also that the behaviour of the task function shown in Figure 5.38a slightly differs from the previous simulation.

The correctness of the final position of the camera is analysed from the results given by Table 5.4. As can be seen, the camera actually gets parallel to the tangent plane Π and the laser depths and the depth Z_P are the expected ones. Furthermore, the secondary task is also fulfilled as can be seen in the final error of the fixation point.

Table 5.4: Final results when positioning with respect to an unknown sphere and a fixation point.

Normal to Π:	$^c(0.0001, -0.0001, 1.0)$
(α_x, α_y) with respect to Π:	$(0.0057°, 0.0040°)$
Z_P :	0.5709 m
Z_P^*:	0.5708 m
Task function $\mathbf{e}$:	$(0.764e - 3, 0.149e - 3, -0.216e - 3)$
Lasers depths:	$(0.6001,\ 0.6001,\ 0.6001,\ 0.6001)$ m
Expected laser depths:	$(0.6,\ 0.6,\ 0.6,\ 0.6)$ m
Fixation point error (pixels):	$(-0.0103, -0.0071)$

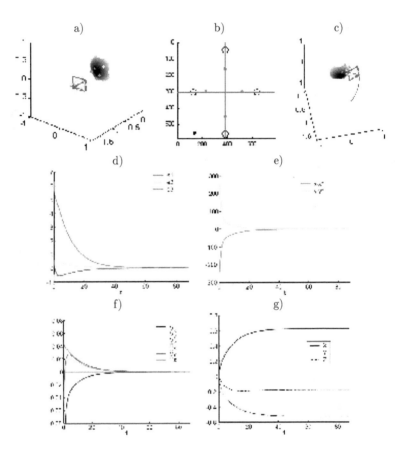

Figure 5.38: Positioning with respect to an unknown sphere and a fixation point. a) Initial position. b) Initial image point distribution (solid red dots) including the laser epipolar lines, the initial position of the fixation point (solid black dot) and the desired image point distribution (circles). c) Camera trajectory till the final position. d) Task function vs. time (in s). e) Secondary task: fixation point error (in pixels). f) Camera velocities (in m/s and rad/s). g) Camera initial position (in m) expressed in the camera frame vs. time (in s).

Third simulation: known sphere

In this case, the conditions described in the first simulation have been adopted. However, it has been assumed that the model of the sphere is known. This allows the real desired visual features to be calculated. The depths of every projected point and the object normals in these points are also available thanks to the object model in the desired state. Concretely, the laser points depths in the desired state are

$$Z_1^* = Z_2^* = Z_3^* = Z_4^* = 0.6402 \ m \tag{5.235}$$

and the normals are

$$\begin{array}{rcl}
(A_1, \ B_1, \ C_1) & = & (0, \ -0.5, \ 0.866) \\
(A_2, \ B_2, \ C_2) & = & (0.5, \ 0, \ 0.866) \\
(A_3, \ B_3, \ C_3) & = & (0, \ 0.5, \ 0.866) \\
(A_4, \ B_4, \ C_4) & = & (-0.5, \ 0, \ 0.866)
\end{array} \tag{5.236}$$

With these parameters, the interaction matrix of the decoupled set of visual features in (5.231) is

$$\mathbf{L_s^*} = \begin{pmatrix} 0 & 0 & -2/L & 0 & 0 & 0 \\ 0 & 7.69 & 0 & -6.93 & 0 & 0 \\ 7.69 & 0 & 0 & 0 & 6.93 & 0 \end{pmatrix} \tag{5.237}$$

By using the real desired visual features and the control law based on the model of interaction matrix above, the simulation results are the ones shown in Figure 5.39. As can be seen, the task function decrease and the camera velocities are smoother as when assuming the object as planar. This improvement is due to the real normal vectors included in the interaction matrix.

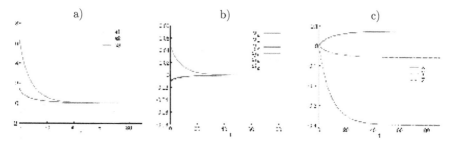

Figure 5.39: Positioning with respect to a known sphere. a) Task function vs. time (in s). b) Camera velocities (in m/s and rad/s). c) Camera initial position (in m) expressed in the camera frame vs. time (in s).

The numeric results of the positioning task are shown in Table 5.5. As can be seen, in this case the camera gets parallel to the tangent plane at the intersecting point between the optical axis and the object at the desired depth $Z^* = 0.6$ m. Note that the final laser depths are also the ones calculated in the desired state.

Table 5.5: Final results when positioning with respect to a known sphere.

Normal to Π:	$^c(0, 0, 1)$
(α_x, α_y) with respect to Π:	$(-0.0284°, -0.0214°)$
Z_P:	0.6 m
Z_P^*:	0.6 m
Task function $\mathbf{e}$:	$(0.0005, -0.0008, 0.0011)$
Lasers depths:	$(0.6292, 0.6291, 0.6293, 0.6293)$ m
Expected laser depths:	$(0.6292, 0.6292, 0.6292, 0.6292)$ m

5.10.3   Case of an elliptic cylinder

An elliptic cylinder is a quadric having one $\lambda_i = 0$ and the other two of the same sign as M. For example, we can express an elliptic cylinder by setting $\lambda_1 = k_1 M$ and $\lambda_3 = k_3 M$ so that

$$k_1 M^o X^2 + k_3 M^o Z^2 - M = 0 \qquad (5.238)$$

with $k_1 > 0$ and $k_3 > 0$. When $k_1 = k_3$ then a *right circular cylinder*, simply known as usual cylinder, is obtained. An example of an elliptic cylinder generated by (5.238) is shown in Figure 5.40.

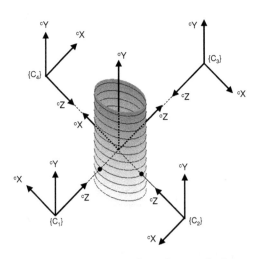

Figure 5.40: Example of an elliptic cylinder.

It is easy to show that there are two types of camera positions for which the task function cancels. Both types of positions are represented in Figure 5.40. The frames $\{C_1\}$ and $\{C_3\}$ represents the cases when the camera is positioned with respect to the minimum curvature regions of the elliptic cylinder. On the other hand, $\{C_2\}$ and $\{C_4\}$ represents camera positions where it is positioned with respect to the maximum curvature regions. Furthermore, there are infinite positions where the task function is 0 as these cases are valid for any position along the ^oY axis.

The case described by $\{C_1\}$ has already been demonstrated in Section 5.10.1 for a generic quadric. The demonstration for $\{C_3\}$ is straightforward as it is a symmetric case.

The demonstration for the camera position $\{C_2\}$ is as follows. According to Figure 5.40 the frame transformation from $\{C_2\}$ to $\{O\}$ is defined by

$$^o\mathbf{X} = \begin{pmatrix} 0 & 0 & 1 \\ 0 & 1 & 0 \\ -1 & 0 & 0 \end{pmatrix} {}^c\mathbf{X} + \begin{pmatrix} -O_x \\ 0 \\ 0 \end{pmatrix} \tag{5.239}$$

with $O_x > 0$. Therefore, the quadric expressed in the camera frame $\{C_2\}$ is

$$k_3 M X^2 + k_1 M (Z - O_x)^2 - M = 0 \tag{5.240}$$

The expression of the depth closest to the camera origin is

$$Z = O_x - \frac{1}{k_1}\sqrt{k_1(1 - k_3 X^2)} \tag{5.241}$$

which provides the depths of every projected laser point which are

$$Z_1 = Z_3 = O_x - \frac{1}{\sqrt{k_1}}$$

$$Z_2 = Z_4 = O_x - \frac{\sqrt{1 - k_3 L^2}}{\sqrt{k_1}} \tag{5.242}$$

Therefore, the task function components e_2 and e_3 are 0 for this camera position. Furthermore, e_1 is 0 when $Z_1 = Z_3 = Z^*$ which happens when

$$O_x = Z^* + \frac{1}{\sqrt{k_1}} \tag{5.243}$$

The demonstration for the cases represented by $\{C_4\}$ is homologous to this one.

Two simulations showing the task of positioning the camera with respect to an unknown elliptic cylinder are now presented. Then, a simulation taking profit of the cylinder model is shown.

First simulation: elliptic cylinder

An elliptic cylinder of parameters $\lambda_1 = 1$, $\lambda_3 = 0.25$ and $M = 0.09$ has been defined. The initial position of the camera expressed in the object frame is $^o(-0.8, 0, -1.4)$ m. The orientation of the optical axis is defined by $\alpha_x = -25°$ and $\alpha_y = 15°$ which are expressed with respect to the plane $^oZ = 0$. The initial state of the simulation is represented in Figure 5.41a. Note that the lasers are projected between the regions of maximum and minimum curvature of the cylinder. The initial and desired image are plotted in Figure 5.41b. The goal is again to move the camera to a position where the task function cancels. The model of the cylinder is assumed to be unknown. Therefore, it is assumed to be a planar object and the corresponding desired visual features and control law are used. At the end of the simulation the camera reaches the position shown in Figure 5.41c. The plots corresponding to the task function, camera velocities and camera trajectory are shown in Figure 5.41d-f. As can be seen, the task function component most affected by the object curvature is e_2 which controls Ω_x.

The accuracy of the positioning task can be analysed from the numeric results presented in Table 5.6. As can be seen, the final value of e_2 is a higher order than e_3. This causes that the normal to point $\mathbf{P}$ has a final error of about $5°$ in one of the components. We think that the explanation is that the camera was converging to a position of the type $\{C_2\}$ represented in Figure 5.40. However, once the camera has reached the position shown in Figure 5.41c, it stabilises because $Z_1 = Z_3$ and $Z_2 = Z_4$. Nevertheless, the final value of Z_2 and Z_4 is not the one predicted by (5.221) as can be seen in Table 5.6. It means that the final position does not correspond, as clearly seen in Figure 5.41c, to any of the positions shown in Figure 5.40. Therefore, even if the task is cancelled, the predictions presented in Section 5.10.1 are not valid in this case.

This simulation shows that depending on the type of object, there might be more positions where the task function gets cancelled but the normal to point $\mathbf{P}$ does not exactly correspond to the optical axis direction.

Table 5.6: Final results when positioning with respect to an elliptic cylinder.

Normal to Π:	$^c(0.0870, -0.0001, 0.9962)$
(α_x, α_y) with respect to Π:	$(4.992°, -0.0075°)$
Z_P:	0.6 m
Z_P^*:	0.6 m
Task function e:	$(0.0054, 0.00004, -0.0000)$
Lasers depths:	$(0.6004, 0.6447, 0.6004, 0.6447)$ m
Expected laser depths:	$(0.6, 0.6804, 0.6, 0.6804)$ m

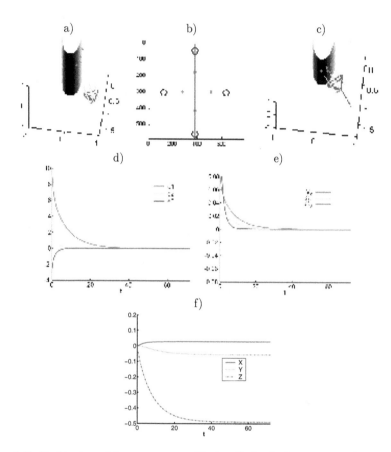

Figure 5.41: Positioning with respect to an elliptic cylinder (only a portion of the object is represented). a) Initial position. b) Initial image point distribution (solid dots) including the laser epipolar lines and the desired image point distribution (circles). c) Camera trajectory till the final position. d) Task function vs. time (in s). e) Camera velocities (in m/s and rad/s). f) Camera initial position (in m) expressed in the camera frame vs. time (in s).

Second simulation: elliptic cylinder including a fixation point

In the second simulation a static point of the object surface has been included in the control scheme. The aim is again to centre such point in the image by using a secondary task. The initial conditions are the same as in the previous simulation. Figure 5.42a shows the initial camera position and the fixation point over the quadric. As can be seen, the aim is to force the camera to get positioned with respect to the zone of maximum curvature as the fixation point lies onto the plane of symmetry $^oX = 0$. The initial and desired images are plotted in Figure 5.42b. The camera converges to the desired position according to the trajectory shown in Figure 5.42c. The curves obtained during the simulation are presented in Figure 5.42d-g.

The numeric results shown in Table 5.7 point out that the fixation point is able to force the camera to reach the type of position denoted as $\{C_2\}$ and shown in Figure 5.40. As can be seen, the normal of the tangent plane and α_x and α_y are better than in the previous example. The task function final residual is also lower. Furthermore, the final laser depths as well as the depth to point $\mathbf{P}$ are the ones predicted by the equations in Section 5.10.1.

Table 5.7: Final results when positioning with respect to an elliptic cylinder and a fixation point.

Normal to Π:	$^c(0.0105,\ 0.0000,\ 0.9999)$
(α_x, α_y) with respect to Π:	$(0.5995°, 0.0023°)$
Z_P:	0.6 m
Z_P^*:	0.6 m
Task function $\mathbf{e}$:	$(0.7082e - 3,\ 0.0809e - 3,\ 0.1241e - 3)$
Lasers depths:	$(0.6001,\ 0.6802,\ 0.6000,\ 0.6803)$ m
Expected laser depths:	$(0.6,\ 0.6804,\ 0.6,\ 0.6804)\ m$
Fixation point error (pixels):	$(0.0064, 0.0040)$

Third simulation: known elliptic cylinder

In this case, the model of the elliptic cylinder is known. This allows the desired position to be defined and calculate the desired visual features, the laser depths and the object normal in every projected point. The desired position denoted as $\{C_2\}$ in Figure 5.40 has been chosen. The desired laser depths which allow the camera to be at a depth $Z_P^* = 0.6$ m of point $\mathbf{P}$ are

$$\begin{aligned} Z_1^* = Z_3^* &= \quad 0.6\ m \\ Z_2^* = Z_4^* &= \quad 0.6804\ m \end{aligned} \tag{5.244}$$

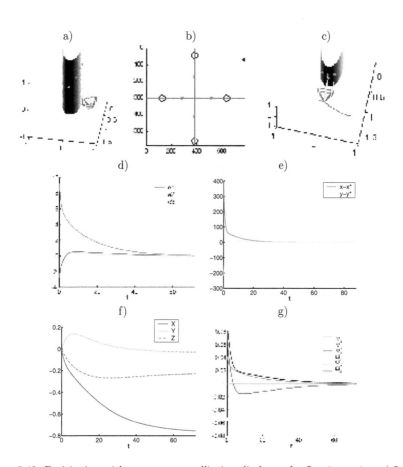

Figure 5.42: Positioning with respect to an elliptic cylinder and a fixation point. a) Initial position. b) Initial image point distribution (solid dots) including the laser epipolar lines, the initial position of the fixation point (solid black dot) and the desired image point distribution (circles). c) Camera trajectory till the final position. d) Task function vs. time (in s). e) Secondary task: fixation point error (in pixels). f) Camera velocities (in m/s and rad/s). g) Camera initial position (in m) expressed in the camera frame vs. time (in s).

and the object normals to every laser point are

$$
\begin{aligned}
(A_1,\ B_1,\ C_1) &= & (0,\ 0,\ 1) \\
(A_2,\ B_2,\ C_2) &= & (0.756,\ 0,\ 0.654) \\
(A_3,\ B_3,\ C_3) &= & (0,\ 0,\ 1) \\
(A_4,\ B_4,\ C_4) &= & (-0.756,\ 0,\ 0.654)
\end{aligned}
\tag{5.245}
$$

With these parameters, the interaction matrix of the decoupled set of visual features in (5.231) is

$$
\mathbf{L_s^*} = \begin{pmatrix}
0 & 0 & -2/L & 0 & 0 & 0 \\
0 & 0 & 0 & -2 & 0 & 0 \\
15.39 & 0 & 0 & 0 & 12.47 & 0
\end{pmatrix}
\tag{5.246}
$$

By using the real desired visual features and the control law based on the model of interaction matrix above, the simulation results are the ones shown in Figure 5.43. The final position of the camera is quite similar to the one obtained in the first simulation. Therefore, the camera is not able to reach the desired position.

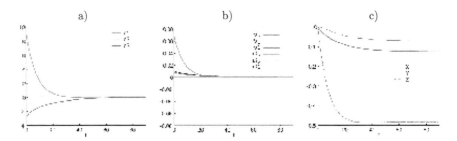

Figure 5.43: Positioning with respect to a known elliptic cylinder. a) Task function vs. time (in s). b) Camera velocities (in m/s and rad/s). c) Camera initial position (in m) expressed in the camera frame vs. time (in s).

The reason is found in Table 5.8. As can be seen, the position reached by the camera also cancels the task function even if the laser depths are not the desired ones. Note that the normal vector to point $\mathbf{P}$ is worst in this case as shown by the error in α_x. This example shows that taking profit of the object model for calculating the desired visual features and making more complex the interaction matrix does not always improves the results. This is because there are multiple positions different to the desired one where the task function cancels.

Table 5.8: Final results when positioning with respect to a known sphere.

Normal to Π:	$^c(0.1051,\ -0.0004,\ 0.9945)$
(α_x, α_y) with respect to Π:	$(6.0310°,\ 0.0230°)$
Z_P:	0.5996 m
Z_P^*:	0.6 m
Task function e:	$(-0.0024,\ 0.0008,\ -0.0090)$
Lasers depths:	$(0.5999,\ 0.6535,\ 0.5998,\ 0.6521)$ m
Desired laser depths:	$(0.6,\ 0.6804,\ 0.6,\ 0.6804)\ m$

5.10.4   Case of a hyperbolic cylinder

A hyperbolic cylinder is a quadric having one $\lambda_i = 0$ and the other two of different sign. For example, it can noted by setting $\lambda_3 = -k_3\lambda_1$ and

$$\lambda_1{}^oX^2 - k_3\lambda_1{}^oZ^2 - M = 0 \qquad (5.247)$$

with $k_3 > 0$. The type of hyperbolic cylinder generated by this equation is shown in Figure 5.44.

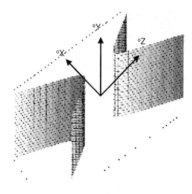

Figure 5.44: Example of hyperbolic cylinder.

First simulation: hyperbolic cylinder

A hyperbolic cylinder of parameters $\lambda_1 = -3$, $\lambda_3 = 1$ and $M = 0.09$ has been defined. The initial position of the camera expressed in the object frame is $^o(0, 0, -1.4)$ m. The orientation is defined by $\alpha_x = -2°$ and $\alpha_y = 15°$ which are expressed with respect to the plane $^oZ = 0$. The initial state of the simulation is represented in Figure 5.45a. The initial and desired image are plotted in Figure 5.45b. At the end of the simulation the camera reaches the position shown in Figure 5.45c. As can be seen, the camera converges to the closest region with minimum curvature. The plots corresponding to the task function, camera velocities and camera trajectory are shown in Figure 5.45d-f.

The numeric results corresponding to the final position are presented in Table 5.9. As can be seen, the final position of the simulation is pretty accurate as shown by the normal to point **P** and the low task function error, as well as the final depth. This is possible because the low curvature regions of the hyperbolic cylinder are almost planar.

Other simulations have stated that the camera only stabilises in the region of maximum curvature when the initial orientation of the camera is defined by $\alpha_x = 0$. Note that this position is the one for which Section 5.10.1 demonstrates that the task function is 0. However, this simulation shows that this position is not attractive and the camera tends to another final position.

Table 5.9: Final results when positioning with respect to an hyperbolic cylinder.

Normal to Π:	$^c(0.0016,\ -0.0001,\ 1.000)$
(α_x, α_y) with respect to Π:	$(0.0945°,\ 0.057°)$
Z_P:	0.5998 m
Task function **e**:	$(-0.0031,\ 0.0002,\ -0.0023)$
Lasers depths:	$(0.5998,\ 0.5994,\ 0.5998,\ 0.5991)$ m

Second simulation: hyperbolic cylinder including a fixation point

In this case, the camera initial position is still $^o(0, 0, -1.4)$ m but its initial orientation with respect to $^oZ = 0$ is defined by $\alpha_x = -18°$ and $\alpha_y = 10°$. A static point of the quadric belonging to the plane of symmetry $^oX = 0$ has been used as fixation point as shown in Figure 5.46a. Then, the objective is that the camera gets positioned with respect to the maximum curvature region of the hyperbolic cylinder as in the case studied in Section 5.10.1. The initial and desired images are shown in Figure 5.46b. The camera trajectory is shown in Figure 5.46c. The camera velocities and the trajectory are plotted in Figure 5.42d-g. As can be seen, the system response shows few oscillations which progressively weaken until reaching the convergence.

The information concerning the final position is shown in Table 5.10. As can be seen, the approach based on the fixation point allows the desired position to be reached. Note that the task function is cancelled and the laser depths and Z_P correspond to the ones predicted by (5.221) and (5.225), respectively, shown in Table 5.10.

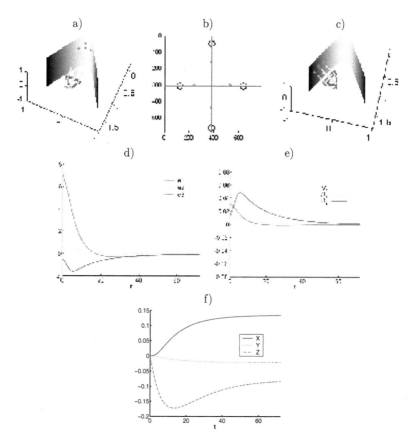

Figure 5.45: Positioning with respect to an hyperbolic cylinder. a) Initial position. b) Initial image point distribution (solid dots) including the laser epipolar lines and the desired image point distribution (circles). c) Camera trajectory till the final position. d) Task function vs. time (in s). e) Camera velocities (in m/s and rad/s). f) Camera initial position (in m) expressed in the camera frame vs. time (in s).

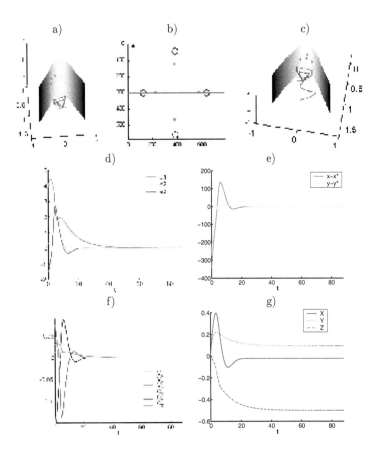

Figure 5.46: Positioning with respect to an hyperbolic cylinder and a fixation point. a) Initial position. b) Initial image point distribution (solid dots) including the laser epipolar lines, the initial position of the fixation point (solid black dot) and the desired image point distribution (circles). c) Camera trajectory till the final position. d) Task function vs. time (in s). e) Secondary task: fixation point error (in pixels). f) Camera velocities (in m/s and rad/s). g) Camera initial position (in m) expressed in the camera frame vs. time (in s).

Table 5.10: Final results when positioning with respect to an hyperbolic cylinder and a fixation point.

Normal to Π:	$^c(-0.0318, 0.0, 0.9995)$
(α_x, α_y) with respect to Π:	$(-0.182°, 0.012°)$
Z_P:	0.600 m
Z_P^*:	0.6 m
Task function **e**:	$(0.000745, 0.0000510, -0.0000071)$
Lasers depths:	$(0.6000, 0.5033, 0.6000, 0.5033)$ m
Desired laser depths:	$(0.6, 0.5031, 0.6, 0.5031)$ m
Fixation point error (pixels):	$(-2.23e - 4, 0.0021)$

5.10.5   Experiments with a non-planar object

This section presents two experiments where a non-planar object similar to an elliptic cylinder has been used. Concretely, the object already presented in Figure 4.12 has been used. As the exact model of the object is unknown no numeric results concerning the accuracy of the positioning tasks are available. With these examples we want to show that the decoupled image-based approach is able to converge even if the object has strong curvature.

First experiment

In this example, the camera has been roughly positioned in front of the maximum curvature region of the object at a distance of about $^cZ = 90$ cm. Then, a rotation of $20°$ about the ^cY axis of the camera has been applied. The initial image perceived in this state is shown in Figure 5.47a. The visual features obtained from the desired image learned with a plane positioned at $Z^* = 60$ cm have been used in the control law. The task function and the camera velocities obtained are plotted in Figure 5.47c-d. As can be seen, the task function shows a nice decrease to 0, as well as the camera velocities. Note that the larger rotational error appears in e_3 which must cancel the large rotation used for defining the initial position of the camera. The final image obtained in the experiment is shown in Figure 5.47b.

Further experiments have shown that the system is able to converge whenever the lasers are projected onto the object in the initial state.

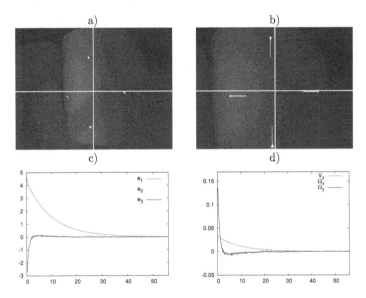

Figure 5.47: Experiment with a non-planar object. a) Initial image. b) Final image containing the lasers point traces. c) Task function vs. time (in s). d) Camera velocities (in m/s and rad/s).

Second experiment

In this example, a fixation point has been used. Therefore, this experiment shows the behaviour of the system when a secondary task aiming to centre the point in the image is simultaneously used. Then, the goal is to get parallel to the fixation point at a distance of $Z^* = 60$ cm.

The initial image of the experiment is shown in Figure 5.48a. The fixation point is the black spot. The image corresponding to the end of the experiment is presented in Figure 5.48b, while the task function and the camera velocities are plotted in Figure 5.48c-d. Note that this experiment shows that it is possible to position the camera with respect to a single point of a quadric object by using the laser points. However, the orientation error with respect to the tangent plane in the fixation point is unknown.

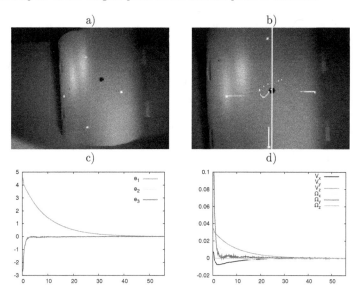

Figure 5.48: Experiment with a non-planar object and a fixation point. a) Initial image (the fixation point is the black spot). b) Final image containing the lasers point and fixation point traces. c) Task function vs. time (in s). d) Camera velocities (in m/s and rad/s).

5.11   Conclusions

This chapter has presented a solution to the classic plane-to-plane positioning task from the combination of visual servoing and structured light. The projection of structured light not only simplifies the image processing but also allows the system to deal with low-textured objects lacking of visual features. A structured light sensor for eye-in-hand systems has been proposed. The sensor is based on four laser pointers attached to a cross-structure. Such a configuration has been chosen in order to obtain an optimal distribution of image points which is invariant to depth once the camera is parallel to the object. A position-based approach and several image-based approaches have been presented. The former is based on reconstructing the parameters of the plane equation of the object and has shown pretty decoupling and robustness against calibration errors. However, it requires to robustly reconstruct the parameters by solving a system of non-linear equations at each iteration. This process may be sensible to image noise depending on the numeric algorithm used. On the other hand, the image-based approaches have been analytically compared through stability analysis in front of different types of misalignments and camera calibration errors.

Two of the image-based approaches have shown a pretty robustness against calibration errors. The first is based on the area and several angles extracted from the polygon containing the four points in the image. Such features have been normalised in order to obtain a linear mapping from the task function space to the camera velocities near the desired state. Furthermore, they exhibit a nice decoupling near the desired state. The good performance of these features has been experimentally demonstrated. However, the high complexity of the analytic expressions of the features has avoided to obtain analytic results concerning either the global asymptotic stability or the camera trajectory generated by the constant control law based on the interaction matrix evaluated in the desired state.

On the other hand, a set of visual features based on non-complex non-linear combinations of the image point coordinates has also obtained very good performance. The advantage of these features is that they decouple rotational from translational dof in all the workspace (not only around the desired state as in the previous approach). This decoupling is possible because these image-based features are proportional to the object plane parameters used in the position-based approach. By evaluating the interaction matrix for the desired state it can be seen that these features also produce a linear mapping from the task function space to the camera velocities around the desired state, without needing any normalisation. Thanks to the decoupled form of the general interaction matrix, it has been possible to prove the global asymptotic stability under ideal conditions (so that it is sure that the system converges for any initial camera-object relative pose where the visibility constraint holds), and the analytic expression of the camera trajectory in the space. The advantage of this control law is that both the task function and the camera velocities are monotonic producing good camera trajectories. The main drawback of this approach is its sensitivity to large calibration errors like a large misalignment between the camera and the laser-cross. However, its robustness has been improved (the analytic proof has been provided) by defining a corrected version of the features based on a planar transformation applied to the image. The corrected visual features have obtained the same robustness against large laser-cross misalignment that the features based on the area and the angles. The camera velocities produced by these two techniques in the simulations are almost the same, since both sets of visual features are proportional when the camera is nearly parallel to the object. In the real experiments under large calibration errors similar performances have been obtained.

An interesting characteristic of the image-based approach based on the decoupled features and the position-based approach is that the interaction matrix can be estimated at each iteration from the feature vector. Hence, a non-constant law based on the estimated interaction matrix can be also used, obtaining a camera trajectory very similar to a straight line even in presence of small calibration errors. The image-based approach has the advantage that the feature vector is calculated only from image data and, unlike the position-based approach, it does not require to solve a non-linear system of equations. Experiments have shown that in presence of large calibration errors the robot is not always capable of reaching the desired position since the velocities produced by the control law become strongly non-linear. Therefore, it seems preferable to use the constant control law which produces almost strictly monotonic velocities in presence of large calibration errors.

The behaviour of the decoupled image-based approach in presence of non-planar objects has also been studied. Concretely, quadric objects have been considered. Firstly, the case of quadric objects of unknown model has been addressed. In this case, the object is assumed to be planar in order to define the desired visual features and the control law. It has been analytically proven that there exist, at least, a position of the camera for which the task function is cancelled. Furthermore, in this position the camera gets parallel to the tangent plane of the intersection point between the optical axis and the object. Simulations have shown, however, that in some cases, depending on the object, the task can get cancelled in other positions where the accuracy of the positioning task is not so good. Other simulations show that it is possible to force the camera to reach a given desired position by using a unique fixation point corresponding to a physical point of the object's surface. Experimental results with a non-planar object validate the approach. On the other hand, the case when the model of the quadric object is available has also been addressed. In this case, the desired position can be analytically defined obtaining the real desired visual features. Furthermore, information about the object curvature in the desired state can be included in the interaction matrix. Simulations have shown that in some cases this improves the results. However, the decoupling of the interaction matrix is lost and the improvement is not ensured as shown by other simulations.

Finally, we remark that the level of decoupling achieved in this work is due to the fact that the points are projected. Such decoupling has not been reached with visual features extracted from the object itself. We have also shown that an appropriate choice of the light pattern can be used for optimising the control law.

Chapter 6

Conclusions and further work

This chapter presents the conclusions and some perspectives opened by this work. The scientific contributions of the work are first discussed. Afterwards, the list of publications related to this work is presented as well as the scientific collaborations involved during its preparation. Finally, further work and future perspectives are discussed.

6.1 Conclusions

This thesis has focused on the combination of visual servoing and structured light for positioning a robot with respect to objects observed by a camera. Most part of techniques in visual servoing rely on extracting visual features from the objects of interest. However, these techniques are valid as long as the objects are textured, and good lighting conditions are available, or if they have artificial landmarks. Therefore, there is a lack of approaches to visual servoing which are able to deal with positioning tasks with respect to non-textured objects or under adverse lighting conditions. Furthermore, every visual servoing technique is dependent to the object appearance, which can be pretty complex in case of natural objects.

The solution pointed out in this thesis is based on using structured light emitters for projecting light patterns onto the objects. In this manner, visual features are always available independently of the object appearance. Furthermore, if the patterns are encoded, correspondences between images taken under different points of view are robustly and unambiguously found. Another research topic that has been treated is how to optimise the control law in visual servoing thanks to the flexibility of structured light. In this case, the projected pattern can be designed in order to optimise the control law in terms of decoupling, stability and robustness against calibration errors. We remark that visual servoing based on visual features provided by structured lighting is still a very unexplored research area. Therefore, this thesis intends to investigate and contribute into this field.

Firstly, a comprehensive study on coding strategies for structured light has been presented. Coded structured light is typically used in active stereovision by using a camera and a video-projector placed aside. By taking the coding strategy used into account, robust correspondences are found between the camera image and the projected pattern. Subsequently, if the devices are calibrated, the illuminated object can be reconstructed by triangulating the correspondences. This part of the work updates the survey presented in [Batlle *et al.*, 1998] and proposes a novel and more consistent classification of the existing techniques. One of the contributions of this study is the comparative evaluation of a group of representative coded patterns by using a common experimental setup. The results allow the most common patterns to be compared in terms of resolution, i.e. number of correspondences, accuracy and quality of the reconstructed objects. The extensive classification and explanation of each group of techniques bring valuable guidelines for easily deciding which type of pattern must be used depending on the application requirements.

The survey on coded structured light has shown that techniques exploiting time-multiplexing obtain the highest resolution and accuracy since the number of patterns is unrestricted and therefore more information can be projected. However, since a sequence of patterns must be projected, this type of techniques are usually restricted to reconstruct static objects. In many applications this constraint cannot be tolerated because either the object is moving or the camera and the projector are not static. The most suitable alternative for these cases are patterns encoded using spatial neighbourhood. In this cases a unique image must be acquired as a unique pattern is projected. The survey comparison states that the best results in terms of both resolution and accuracy are obtained with stripe and multi-slit patterns. However, in order to obtain a robust coding scheme and fair resolution it is usually necessary to use a considerable number of colours which increases the sensitivity of the pattern in front of colourful objects. As result, a new colour encoded stripe pattern has been proposed which improves the resolution of typical one-shot patterns by using less colours than usual. The new approach has been compared through quantitative and qualitative results to similar patterns in order to validate it. The results show that with the new pattern the resolution is 1.5 times greater than the one obtained with a classic stripe pattern. Furthermore, the number of required colours is reduced to the half. Moreover, the $3D$ reconstruction results show that the sub-pixel accuracy on the determination of the correspondences is more accurate when using the new pattern. However, as the new pattern still projects colour, its performance in presence of colourful objects decreases as in any other similar pattern.

A second important contribution of this thesis is the study of the applicability of coded structured light in a visual servoing framework. Up to our knowledge, there are no previous works considering the projection of a coded pattern for visual servoing purposes. The advantage of this approach is that the existing techniques of visual servoing can be directly applied without any adaptation. Experiments have been carried out with a 6 dof robot with eye-in-hand configuration and a video-projector placed aside the robotic cell. A pattern consisting of an array of coloured dots has been used in the experiments for controlling the robot. The pattern provides robust correspondences among the dots imaged from different points of view. Therefore, finding point correspondences between the desired image and the initial and intermediate images becomes straightforward. A control law based on the normalised coordinates of the image points assuming a constant depth distribution has been used. Even with this simple control law, the robot has been able to reach the desired position in the case of planar and non-planar objects. Nevertheless, the behaviour with the non-planar object has been less satisfactory, probably due to the strong assumptions about the depth distribution of the points used in the control law. The most important conclusion is that any of the classic image-based visual servoing approaches can be used by projecting a suitable coded pattern. The advantage of using coded structured light is that the system becomes independent to the object appearance. This allows to deal not only with non-textured objects but also with objects too complex for robustly extracting and tracking visual features. However, two main drawbacks have been identified. Firstly, this approach is only valid when the desired image is known, as in classic image-based visual servoing, and when the object remains in the field of view of the static projector. Secondly, the number of existing coded structured light patterns which can be used for controlling the 6 degrees of freedom is quite reduced. This is due to that for visual servoing applications it is necessary that patterns are rotation invariant.

Finally, another potentiality of using structured light has been addressed. Concretely, the ability of optimising the control law thanks to a specific design of the projected pattern has been investigated. For this purpose, the classic plane-to-plane positioning task has been treated when using an eye-in-hand configuration. In this case, the dedicated structured light has been placed onboard so that it remains linked to the camera. The structured light emitter proposed for this task consists of four low-cost laser pointers attached to a cross centred in the camera frame. The first advantage of this configuration is that the laser points distribution in the image when the camera is parallel to the object plane at the desired depth is valid for any planar object independently of its texture. A position-based and several image-based visual servoing approaches based on the 4 points provided by the lasers have been formulated. The position-based approach has shown good performance even if no analytic results concerning its stability have been obtained. However, it requires a non-linear optimisation step at each iteration for robustly estimating the object parameters. On the other hand, a decoupled image-based approach based on a non-linear combination of the image point coordinates has obtained great results. The visual features decouple the rotational part of the interaction matrix from the translational one for any camera-object relative pose. Thanks to the decoupled form of the interaction matrix, the global asymptotic stability under ideal conditions has been proven. In addition to this, the robustness against misalignments between the camera and the structured light emitter has been proven through the local asymptotic stability and experimental results. Therefore, the main contribution of this part of the work is to provide a robust control law thanks to a specific design of the structured light sensor. This type of results had never been obtained with a visual servoing approach based on visual features extracted from the object itself. As a drawback, we can mention that the validity of the proposed sensor is ensured if the four lasers point towards the same direction. Such a perfect alignment is pretty difficult to achieve in real conditions. Nevertheless, experiments show that good results are achieved even when the lasers have slightly different orientations. We have also studied what happens when the decoupled image-based approach is used when positioning the camera with respect to non-planar objects of known and unknown model. Concretely, quadric objects have been considered. Analytic and simulation results show that it is possible to cancel the task function with respect to these objects so that the camera gets parallel to the tangent plane of the object in certain points. Furthermore, if the object model is known, some information about the object's shape can be included in the interaction matrix. This can improve the positioning task in some cases. However, we remark that the decoupling achieved for the case of planar objects is lost. Nevertheless, it seems that an appropriate orientation of the lasers taking into account the shape of the object might lead to a partially decoupled interaction matrix.

Overall, the combination of structured light and visual servoing has been proved to have a great potentiality for enlarging the application field on visual control. Indeed, structured light and coded light patterns provide a rich variety of visual features which can be specifically designed in order to optimise the control law.

6.2   Perspectives

The domain of visual servoing based on structured light rests mainly unexplored. Therefore, the present thesis requires some further work and opens several perspectives of future research.

First of all, some further work based on the proposed approach should be done in order to solve some problems that have appeared. For example, during the experiments of robot positioning by using points provided by the deported video-projector, we have realised that a large number of points must be matched from the initial to the desired image in order to success. It seems that the problem is related to the distribution of the points in the desired image, which has an important repercussion on the conditioning of the interaction matrix. As made in [Feddema *et al.*, 1991], an optimal choice of the image points should be studied in order to improve the matrix conditioning. Nevertheless, the problem should be studied in depth in order to explain why a large number of correspondences are required specially when the object is non-planar. In the case of planar objects it is known that the control law is able to converge by using only 4 points. Furthermore, it is also interesting to study the influence of taking a variable number of points at each iteration, because it allows occlusions to be treated. Alternatively, other visual features should be used for improving the control law like the image moments of the point distribution [Tahri and Chaumette, 2004] or the extended visual servoing [Schramm *et al.*, 2004]. Another option that should be studied is the use of the point depths obtained by triangulating correspondences between the camera image and the projected pattern. This is easy to do as the correspondence problem is easily solved in our case.

Apart from this further work, the present thesis opens several perspectives of future research.

Firstly, some perspectives for coded structured light are proposed. In this field, the use of colour to get a large number of correspondences from a unique pattern is very extended. However, in presence of highly saturated colourful objects, difficulties when distinguishing all the projected colours appear. The current tendency is to define sophisticated grey-level patterns for increasing the robustness against the object colours. However, since a unique pattern is projected, either the accuracy decreases or a non-absolute coding scheme is applied. The latter means that only several elements of the pattern are uniquely encoded and the rest must be decoded by using propagation of neighbouring constraints. There remains therefore the goal of designing an absolute-coded pattern using the minimum number of grey levels. In this direction, a future work consists of studying the adaptation of the pattern proposed in Chapter 3 to the case of grey-levels.

This thesis has presented a first step towards the use of coded light patterns in a visual servoing framework. In the case of using a deported video-projector, several open issues remain. Firstly, it is necessary to the design a new coding strategy which is able to generate patterns being rotation invariant. These patterns will be very useful for visual servoing purposes. On the other hand, it is interesting to identify which primitives must contain the pattern in order to optimise the control law. It is well known that in classic image-based visual servoing the use of some primitives can lead to singularities in the interaction matrix or bad conditioning [Chaumette, 1998]. Avoiding this type of problems by using an appropriate pattern design is another advantage of the flexibility of coded structured light. Another goal is to specifically design the pattern in order to optimise the control law for specific tasks like it has been done for the case of the plane-to-plane positioning task. The most ambitious perspective is to take profit of the projected pattern for performing robust positioning tasks with respect to unknown non-planar objects. It is well known that image-based approaches have problems of convergence with non-planar objects when the depth distribution is not well estimated [Malis and Rives, 2003]. Therefore, the case of non-planar objects is still matter of research. In our opinion, it seems feasible to use patterns projecting $2D$ contours onto non-planar objects. $2D$ contours have been already used in monocular vision for obtaining local surface reconstruction as in [Ulupinar and Nevatia, 1993]. This type of techniques belong to the group of *shape from contours or silhouettes* [Cipolla and Blake, 1992]. Contours have already been used in visual servoing for the case of planar objects [Colombo and Allotta, 1999; Drummond and Cipolla, 1999; Collewet and Chaumette, 2000]. The projected pattern could be designed in order to provide the required $2D$ contours for obtaining the $3D$ information needed for positioning the camera with respect to an unknown object. However, a more interesting alternative would be to design an image-based visual servoing approach based on the deformation parameters of these contours in the image as long as the robot moves.

The case of an onboard structured light emitter also presents many perspectives. Indeed, this type of configuration has to be considered since it is the one that must be chosen in case of mobile robots. The first perspective is the use of non heavy projectors being able to project encoded patterns. Nowadays, the dimensions, the weight and the cost of video-projectors are being progressively reduced. Therefore, it is likely to predict that in few years such type of devices will be suitable to be integrated in a robot end-effector or in mobile robots. This will give a large flexibility when designing patterns for robot positioning. Nevertheless, the use of structured light sensors based on lasers must be still considered as a lower cost option. Therefore, different types of laser primitives other than points like planes, circles or grids should be studied for performing more complex positioning tasks or taking non-planar objects into account. In this direction, the structured light emitter could be specially designed in order to produce a robust control law similarly to what it has been done in Chapter 5 for the case of plane-to-plane positioning. An important future work is, as already mentioned for the case of a deported projector, performing positioning tasks with respect to non-planar objects. In this case, since the relative pose between the camera and the structured light emitter remains fixed, $3D$ information of the object can be obtained by calibrating these devices. However, it seems more interesting to consider an uncalibrated or a coarsely calibrated structured light sensor and developing image-based visual servoing approaches based on the deformations of the projected pattern which provide cues of the object curvature. This is the main unexplored area of assisted visual servoing by means of structured light and opens an exciting research direction which is expected to enlarge its application field.

Appendix A

Interaction matrix of μ

Samson et al. [Samson *et al.*, 1991] modelled a general thin-field rangefinder which obtains a measure of the depth to an object along a straight line. According to our notation, the case studied by Samson et al. is represented in Fig. A.1.

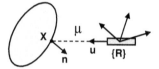

Figure A.1: Thin-field rangefinder schema

Note that $\underline{n}$ is the normal to the object at point $\mathbf{X}$. The measure of the depth that the sensor obtains is expressed in the rangefinder frame denoted as $\{R\}$. From the variation of the distance μ due to the sensor motion found by Samson et al., we can extract the following interaction matrix

$$^{R}\mathbf{L}_{\mu} = -\frac{1}{\underline{n}^{\top}\underline{u}} \left(\underline{n}^{\top} \mid \mu \left(\underline{u} \times \underline{n} \right)^{\top} \right) \tag{A.1}$$

where both $\underline{n}$ and $\underline{u}$ are expressed in the sensor frame.

In our case, when the system is composed by a camera and a laser pointer as shown in Fig. A.2, both $\underline{n}$ and $\underline{u}$ are expressed in the camera frame, as well as the interaction matrix of μ.

As shown in (5.13), the interaction matrix of μ expressed in the camera frame is

$$^{C}\mathbf{L}_{\mu} = -\frac{1}{\underline{n}^{\top}\underline{u}} \left(\mathbf{n}^{\top} \mid \left(\mathbf{X} \times \underline{n} \right)^{\top} \right) \tag{A.2}$$

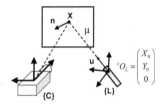

Figure A.2: Our camera-laser system modelling

In order to demonstrate the equivalence with the interaction matrix by Samson et al. it is necessary to express $\mathbf{X}$, $\underline{\mathbf{n}}$ and $\underline{\mathbf{u}}$ in a laser frame with origin equal to $\mathbf{X}_0$. Then from (5.4) we have that point $\mathbf{X}$ in the laser frame is

$$\mathbf{X} = \mu \underline{\mathbf{u}} \tag{A.3}$$

so that (A.2) expressed in the laser frame becomes the same than the interaction matrix by Samson et al.

Andreff et al. also formulated the interaction matrix of μ [Andreff *et al.*, 2002]. In their case, the laser frame was chosen so that the Z axis coincides with the laser direction as shown in Fig. A.3.

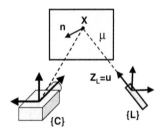

Figure A.3: Camera-laser system modelling by Andreff et al.

The interaction matrix presented by Andreff et al. was expressed in function of α_x and α_y. The former is the angle between Z_L and $\mathbf{n}_x = (A, 0, C)$, being $\mathbf{n}_x$ the projection of $\underline{\mathbf{n}}$ to the plane $Y_L = 0$. Similarly, α_y is the angle between Z_L and $\mathbf{n}_y = (0, B, C)$ which is the projection of $\underline{\mathbf{n}}$ to the plane $X_L = 0$. The geometric interpretation of α_x and α_y is shown in Fig. A.4. Taking into account the sign conventions of the angles and the constraint $C > 0$, we have that

$$
\begin{aligned}
A &= C \tan \alpha_x \\
B &= -C \tan \alpha_y
\end{aligned}
\tag{A.4}
$$

so that expressing $C = \sqrt{1 - A^2 - B^2}$ the following relationship arise

$$
\begin{aligned}
A &= \frac{\tan \alpha_x}{\sqrt{\tan^2 \alpha_x + \tan^2 \alpha_y + 1}} \\
B &= -\frac{\tan \alpha_y}{\sqrt{\tan^2 \alpha_x + \tan^2 \alpha_y + 1}}
\end{aligned}
\tag{A.5}
$$

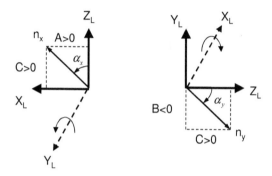

Figure A.4: Geometric interpretation of α_x and α_y when they are positive.

The interaction matrix by Andreff et al. can be derived from the one by Samson et al. in (A.1). If the parameters of this interaction matrix are expressed in the laser frame proposed by Andreff et al. we have that $\underline{u} = (0, 0, 1)$ and

$$
\begin{aligned}
\underline{u} \times \underline{n} &= \begin{vmatrix} \mathbf{i} & \mathbf{j} & \mathbf{k} \\ 0 & 0 & 1 \\ A & B & C \end{vmatrix} = \begin{pmatrix} -B \\ A \\ 0 \end{pmatrix} \\
\mathbf{n}^\top \underline{u} &= C
\end{aligned}
\tag{A.6}
$$

Then, the formula by Samson becomes

$$
\mathbf{L}_\mu = \begin{pmatrix} -A/C & -B/C & -1 & \mu B/C & -\mu A/C & 0 \end{pmatrix}
\tag{A.7}
$$

so that taking into account the relationships in (A.4) we obtain

$$
\mathbf{L}_\mu = \begin{pmatrix} -\tan \alpha_x & \tan \alpha_y & -1 & -\mu \tan \alpha_y & \mu \tan \alpha_x & 0 \end{pmatrix}
\tag{A.8}
$$

which is the form of the interaction matrix proposed by Andreff et al. [Andreff *et al.*, 2002].

Appendix B

Model considering laser-cross misalignment

This appendix presents the model parameters when the laser-cross center is displaced from the camera origin and the orientation of the laser-cross frame is not the same that the camera frame. The camera intrinsic parameters are supposed to be perfectly calibrated. The laser-cross misalignment is modelled according to a homogeneous frame transformation matrix of the form

$$
{}^{C}\mathbf{M}_L = \begin{pmatrix} {}^{C}\mathbf{R}_L & {}^{C}\mathbf{T}_L \\ \mathbf{0}_3 & 1 \end{pmatrix}
\tag{B.1}
$$

which passes from points expressed in the structured light sensor frame to the camera frame (see Fig. 5.4).

First of all, the orientation vector and the reference point of every laser in the camera frame must be calculated taking into account the misalignment. We start from the values of these parameters in the laser-cross frame, which coincide with the ideal parameters shown in Table 5.1. Then, in the camera frame we have that for every laser

$$
{}^{C}\underline{\mathbf{u}} = {}^{C}\mathbf{R}_L{}^{L}\underline{\mathbf{u}}
\tag{B.2}
$$
$$
{}^{C}\mathbf{X}_r = {}^{C}\mathbf{R}_L{}^{L}\mathbf{X}_0 + {}^{C}\mathbf{T}_L
\tag{B.3}
$$

where ${}^{L}\underline{\mathbf{u}} = (0, 0, 1)$. Note that ${}^{C}\mathbf{X}_r$ is a point belonging to the laser direction but it is not the reference point of the laser according to our definition. Remember that the reference point ${}^{C}\mathbf{X}_0$ must lie on the plane $Z_C = 0$. The equation of the line corresponding to the laser direction can be expressed in function of ${}^{C}\mathbf{X}_0$ as follows

$$
{}^{C}\mathbf{X}_r = \mu^{C}\underline{\mathbf{u}} + {}^{C}\mathbf{X}_0
\tag{B.4}
$$

so that since we impose that $^{C}Z_0 = 0$ then

$$\mu = \frac{^{C}Z_r}{^{C}u_z} \tag{B.5}$$

The calculation of the $^{C}\mathbf{X}_0$ is then straightforward

$$^{C}\mathbf{X}_0 = -\frac{^{C}Z_r}{^{C}u_z}{^{C}\underline{\mathbf{u}}} + {^{C}\mathbf{X}_r} \tag{B.6}$$

The model parameters taking into account the whole model of misalignment become too complicated. Instead of this, we present the model parameters under individual types of misalignment, namely a simple displacement of the laser-cross with respect to the camera origin, and individual rotations of the laser-cross around the X, Y and Z axis, respectively.

Displacement In this case, the laser-cross frame has the same orientation that the camera frame ($^{C}\mathbf{R}_L = \mathbf{I}_3$), but its origin has been displaced according to the vector

$$^{C}\mathbf{T}_L = (t_x,\ t_y,\ t_z) \tag{B.7}$$

The model parameters are then the ones shown in Table B.1.

Table B.1: Model parameters under a translational misalignment of the laser-cross.

Laser	X_0	Y_0	x	y	Z
1	t_x	$L + t_y$	t_x/Z_1	$(t_y + L)/Z_1$	$-\dfrac{At_x + BL + Bt_y + D}{C}$
2	$-L + t_x$	t_y	$(t_x - L)/Z_2$	t_y/Z_2	$-\dfrac{At_x - AL + Bt_y + D}{C}$
3	t_x	$-L + t_y$	t_x/Z_3	$(t_y - L)/Z_3$	$-\dfrac{At_x - BL + Bt_y + D}{C}$
4	$L + t_x$	t_y	$(t_x + L)/Z_4$	t_y/Z_4	$-\dfrac{At_x + AL + Bt_y + D}{C}$

Rotation around the X_C axis The laser-cross is centered in the camera origin, but the laser-cross frame is rotated a certain angle ψ with respect to the X axis of the camera frame. The rotation is restricted to the interval $\psi \in (-\pi/2, \pi/2)$, otherwise the lasers projection is out of the camera field of view. The rotation matrix $^{C}R_{L}$ is then

$$\text{Rot}(X, \psi) = \begin{pmatrix} 1 & 0 & 0 \\ 0 & c\psi & -s\psi \\ 0 & s\psi & c\psi \end{pmatrix} \tag{B.8}$$

where $c\psi = \cos\psi$ and $s\psi = \sin\psi$. The model parameters under this type of misalignment are shown in Table B.2.

Table B.2: Model parameters when the laser-cross is rotated around X_C.

Laser	X_0	Y_0	x	y	Z
1	0	$\dfrac{L}{c\psi}$	0	$-\dfrac{Ds\psi + LC}{BL + Dc\psi}$	$\dfrac{BL + Dc\psi}{Bs\psi - Cc\psi}$
2	$-L$	0	$-\dfrac{L(Bs\psi - Cc\psi)}{c\psi(D - AL)}$	$-\dfrac{s\psi}{c\psi}$	$\dfrac{c\psi(D - AL)}{Bs\psi - Cc\psi}$
3	0	$-\dfrac{L}{c\psi}$	0	$\dfrac{LC - Ds\psi}{Dc\psi - BL}$	$\dfrac{-BL + Dc\psi}{Bs\psi - Cc\psi}$
4	L	0	$\dfrac{L(Bs\psi - Cc\psi)}{c\psi(AL + D)}$	$-\dfrac{s\psi}{c\psi}$	$\dfrac{c\psi(AL + D)}{Bs\psi - Cc\psi}$

Rotation around the Y_C axis. Let us present now the case where the laser-cross is centered in the camera origin, but it is rotated an angle $\theta \in (-\pi/2, \pi/2)$ with respect to the Y axis of the camera frame. The rotation matrix $^{C}R_{L}$ has the following form

$$\text{Rot}(Y, \theta) = \begin{pmatrix} c\theta & 0 & s\theta \\ 0 & 1 & 0 \\ -s\theta & 0 & c\theta \end{pmatrix} \tag{B.9}$$

where $c\theta = \cos\theta$ and $s\theta = \sin\theta$. The model parameters are then the ones shown in Table B.3.

Rotation around the Z_C axis In case that a rotation of ϕ occurs around the Z axis of the camera frame, the rotation matrix is

$$\text{Rot}(Z, \phi) = \begin{pmatrix} c\phi & -s\phi & 0 \\ s\phi & c\phi & 0 \\ 0 & 0 & 1 \end{pmatrix} \tag{B.10}$$

where $c\phi = \cos\phi$ and $s\phi = \sin\phi$. In this case, the model parameters are shown in Table B.4.

Table B.3: Model parameters when the laser-cross is rotated around Y_C.

Laser	X_0	Y_0	x	y	Z
1	0	L	$\dfrac{s\theta}{c\theta}$	$-\dfrac{L\left(As\theta + Cc\theta\right)}{c\theta\left(BL + D\right)}$	$-\dfrac{c\theta\left(BL + D\right)}{As\theta + Cc\theta}$
2	$-\dfrac{L}{c\theta}$	0	$\dfrac{Ds\theta + LC}{-AL + Dc\theta}$	0	$\dfrac{-AL + Dc\theta}{As\theta + Cc\theta}$
3	0	$-L$	$\dfrac{s\theta}{c\theta}$	$\dfrac{L\left(As\theta + Cc\theta\right)}{c\theta\left(-BL + D\right)}$	$-\dfrac{c\theta\left(-BL + D\right)}{As\theta + Cc\theta}$
4	$\dfrac{L}{c\theta}$	0	$\dfrac{Ds\theta - LC}{AL + Dc\theta}$	0	$-\dfrac{AL + Dc\theta}{As\theta + Cc\theta}$

Table B.4: Model parameters when the laser-cross is rotated around Z_C.

Laser	X_0	Y_0	x	y	Z
1	$-s\phi L$	$c\phi L$	$-\dfrac{Ls\phi}{Z_1}$	$\dfrac{Lc\phi}{Z_1}$	$\dfrac{As\phi L - Bc\phi L - D}{C}$
2	$-c\phi L$	$-s\phi L$	$-\dfrac{Lc\phi}{Z_2}$	$-\dfrac{Ls\phi}{Z_2}$	$\dfrac{Ac\phi L + Bs\phi L - D}{C}$
3	$s\phi L$	$-c\phi L$	$\dfrac{Ls\phi}{Z_3}$	$-\dfrac{Lc\phi}{Z_3}$	$-\dfrac{As\phi L - Bc\phi L + D}{C}$
4	$c\phi L$	$s\phi L$	$\dfrac{Lc\phi}{Z_4}$	$\dfrac{Ls\phi}{Z_4}$	$-\dfrac{Ac\phi L + Bs\phi L + D}{C}$

Appendix C

Kinematic screw frame transformation

The objective of this appendix is to define a frame transformation which allows the kinematic screw typically expressed in the camera frame to be expressed in a frame attached to the object. This can be done by using a transformation like

$$^c\mathbf{v} = {}^c\mathbf{T}_o\,{}^o\mathbf{v} \tag{C.1}$$

where $^c\mathbf{v}$ and $^o\mathbf{v}$ are the kinematic screw expressed in the camera and the object frames, respectively, and $^c\mathbf{T}_o$ is the 6×6 transformation changing the basis frame. This transformation is then useful to express the interaction matrix $\mathbf{L_x}$ in the object frame, which can be used to check which type of object motions can be detected in the camera image by using a certain set of visual features. The time variation of the visual features can be expressed in two ways

$$\dot{\mathbf{s}} = {}^c\mathbf{L_s}\,{}^c\mathbf{v} \tag{C.2}$$

$$\dot{\mathbf{s}} = {}^o\mathbf{L_s}\,{}^o\mathbf{v} \tag{C.3}$$

where $^c\mathbf{L_s}$ and $^o\mathbf{L_s}$ are the interaction matrices expressed in the camera and object frames, respectively. Then, by plugging (C.1) into (C.2), we can write

$$\dot{\mathbf{s}} = {}^c\mathbf{L_s}\,{}^c\mathbf{T}_o\,{}^o\mathbf{v} \tag{C.4}$$

so that according to (C.3) the interaction matrix expressed in the object frame is

$$^o\mathbf{L_s} = {}^c\mathbf{L_s}\,{}^c\mathbf{T}_o \tag{C.5}$$

The transformation matrix has the following form

$$
{}^{c}\mathbf{T}_{o} = \begin{pmatrix} {}^{c}\mathbf{R}_{o} & [{}^{c}\mathbf{P}_{o}]_{\times}{}^{c}\mathbf{R}_{o} \\ \mathbf{0}_{3} & {}^{c}\mathbf{R}_{o} \end{pmatrix}
\tag{C.6}
$$

where ${}^{c}\mathbf{R}_{o}$ is the rotation matrix from the camera frame to the object frame, and ${}^{c}\mathbf{P}_{o}$ is the origin of the frame $\{O\}$ expressed in the object frame $\{C\}$. $[{}^{c}\mathbf{P}_{o}]_{\times}$ is the antisymmetric matrix associated to the vector ${}^{c}\mathbf{P}_{o}$. In Fig. C.1 the frame transformation is represented.

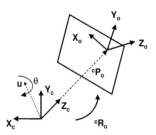

Figure C.1: Frame transformation schema

Let us now present how to obtain the analytic expression of the rotation matrix ${}^{c}\mathbf{R}_{o}$ and the origin of the object frame ${}^{c}\mathbf{P}_{o}$.

C.1   Rotation matrix

The constraint that is fixed to calculate the rotation matrix ${}^{c}\mathbf{R}_{o}$ is that $\mathbf{Z}_{o}$ must be equal to $\mathbf{Z}_{c}$ after applying the rotation (both $\mathbf{Z}_{o}$ and $\mathbf{Z}_{c}$ are expressed in the camera frame). Note that this constraint implies that a single rotation θ is made around an unitary axis $\underline{\mathbf{u}}$ which is orthogonal to $\mathbf{Z}_{c}$ and $\mathbf{Z}_{o}$. The rotation axis $\mathbf{u}$ can be calculated as follows

$$
\mathbf{u} = \mathbf{Z}_{c} \times \mathbf{Z}_{o} = (0,0,1) \times (A,B,C) = (-B,A,0)
\tag{C.7}
$$

Therefore, the unitary vector $\underline{\mathbf{u}}$ is

$$
\underline{\mathbf{u}} = \frac{1}{\sqrt{A^{2}+B^{2}}} \begin{pmatrix} -B \\ A \\ 0 \end{pmatrix}
\tag{C.8}
$$

The rotation θ can be calculated as follows

$$
\begin{aligned}
\sin(\theta) &= \|\mathbf{Z}_{c} \times \mathbf{Z}_{o}\| &\Rightarrow& \quad \sin(\theta) &=& \sqrt{A^{2}+B^{2}} \\
\cos(\theta) &= \mathbf{Z}_{c}^{\top} \cdot \mathbf{Z}_{o} &\Rightarrow& \quad \cos(\theta) &=& C
\end{aligned}
\tag{C.9}
$$

According to the Rodrigues formula, a rotation matrix can be written as

$$^c\mathbf{R}_o = \cos(\theta)\mathbf{I}_3 + \sin(\theta)[\underline{\mathbf{u}}]_\times + (1 - \cos(\theta))\underline{\mathbf{u}} \cdot \underline{\mathbf{u}}^\top \tag{C.10}$$

with

$$[\underline{\mathbf{u}}]_\times = \begin{pmatrix} 0 & -u_z & u_y \\ u_z & 0 & -u_x \\ -u_y & u_x & 0 \end{pmatrix} \tag{C.11}$$

which leads to the rotation matrix

$$^c\mathbf{R}_o = \begin{pmatrix} 1 - \frac{A^2}{1+C} & -\frac{AB}{1+C} & A \\ -\frac{AB}{1+C} & 1 - \frac{B^2}{1+C} & B \\ -A & -B & C \end{pmatrix} \tag{C.12}$$

C.2 Origin of the object frame

The origin of the object frame has been chosen to be equal to the projection of the focal point along the optical axis of the camera onto the planar object. Such a point expressed in the camera frame is

$$^c\mathbf{P}_o = \begin{pmatrix} 0 \\ 0 \\ -D/C \end{pmatrix} \tag{C.13}$$

C.3 Frame transformation

Given the chosen $^c\mathbf{P}_o$ in (C.13) and the calculated rotation matrix $^c\mathbf{R}_o$ in (C.12), the frame transformation $^c\mathbf{T}_o$ has the following expression

$$^c\mathbf{T}_o = \begin{pmatrix} 1 - \frac{A^2}{1+C} & -\frac{AB}{1+C} & A & -\frac{ABD}{C(1+C)} & -\frac{D(B^2-1-C)}{C(1+C)} & \frac{DB}{C} \\ -\frac{AB}{1+C} & 1 - \frac{B^2}{1+C} & B & \frac{D(A^2-1-C)}{C(1+C)} & \frac{ABD}{C(1+C)} & -\frac{AD}{C} \\ -A & -B & C & 0 & 0 & 0 \\ 0 & 0 & 0 & 1 - \frac{A^2}{1+C} & -\frac{AB}{1+C} & A \\ 0 & 0 & 0 & -\frac{AB}{1+C} & 1 - \frac{B^2}{1+C} & B \\ 0 & 0 & 0 & -A & -B & C \end{pmatrix} \tag{C.14}$$

Then, the interaction matrix $\mathbf{L_x}$ expressed in the object frame is

$$^{o}\mathbf{L_x} = {}^{c}\mathbf{L_x} \cdot {}^{c}\mathbf{T}_{o} = \frac{1}{\Pi_0} \begin{pmatrix} 0 & 0 & -\frac{X_0}{Z} & X_0\eta & X_0\xi & 0 \\ 0 & 0 & -\frac{Y_0}{Z} & Y_0\eta & Y_0\xi & 0 \end{pmatrix} \qquad \text{(C.15)}$$

where

$$\eta = \frac{1 - A^2}{C}y + \frac{A(Bx + ACy)}{C(1+C)}$$

$$\xi = \frac{1 - B^2}{C}x + \frac{B(Ay + BCx)}{C(1+C)}$$

$$\Pi_0 = A(X_0 - xZ) + B(Y_0 - yZ) - CZ$$

Therefore, according to $\dot{\mathbf{x}} = {}^{o}\mathbf{L_x}{}^{o}\mathbf{v}$, if the object moved on ${}^{o}V_x$, ${}^{o}V_y$ or ${}^{o}\Omega_z$, the image coordinates of the projected points would not change.

Appendix D

Stability analysis of the decoupled approach

This appendix presents different issues concerning the stability analysis of the set of visual features

$$\mathbf{s} = (\ y_1^{-1} - y_3^{-1}, \ y_1^{-1} + y_3^{-1}, \ x_2^{-1} + x_4^{-1} \ ) \tag{D.1}$$

when using the constant control law based on the interaction matrix evaluated in the desired state $\mathbf{L_s^*}$.

D.1 Solving the differential system

The closed-loop equation of the system corresponding to the set of visual features (5.121) when using the constant control law based on $\mathbf{L_s^*}$ can be written as the following differential system

$$\dot{e}_1(t) = -\frac{\lambda}{4L} \left(e_1(t) \left(4L + e_2(t)^2 L + e_3(t)^2 L \right) + 2Z^* \left(e_2(t)^2 + e_3(t)^2 \right) \right) \tag{D.2}$$

$$\dot{e}_2(t) = -\frac{\lambda}{4} (e_2(t)^3 + 4e_2(t) + e_2(t)e_3(t)^2) \tag{D.3}$$

$$\dot{e}_3(t) = -\frac{\lambda}{4} (e_3(t)^3 + 4e_3(t) + e_3(t)e_2(t)^2) \tag{D.4}$$

The solutions of $e_1(t)$, $e_2(t)$ and $e_3(t)$ can be found as follows. As expected, both $\dot{e}_2(t)$ (D.3) and $\dot{e}_3(t)$ (D.4) are not affected by $e_1(t)$ (D.2). Hence, we can start by searching the solutions of $e_2(t)$ and $e_3(t)$, which control the rotational dof of the system. First, we obtain the expression of $\ddot{e}_2(t)$ by deriving (D.3)

$$\ddot{e}_2(t) = -\frac{\lambda}{4}\left(\dot{e}_2(t)\left(3e_2^2(t) + e_3^2(t) + 4\right) + 2e_2(t)e_3(t)\dot{e}_3(t)\right) \tag{D.5}$$

Afterwards, plugging $\dot{e}_3(t)$ from (D.4) into (D.5)

$$\ddot{e}_2(t) = \frac{\lambda}{8}\left(-\dot{e}_2(t)\left(6e_2^2(t) + 2e_3^2(t) + 8\right) + \right.$$
$$\left. + \lambda e_3^2(t)e_2(t)\left(e_2^2(t) + e_3^2(t) + 4\right)\right) \tag{D.6}$$

From (D.3) the expression of $e_3^2(t)$ can be expressed in function of $\dot{e}_2(t)$ and $e_2(t)$

$$e_3^2(t) = \frac{-4\dot{e}_2(t) - \lambda e_2^3(t) - 4\lambda e_2(t)}{\lambda e_2(t)} \tag{D.7}$$

then, by plugging it into (D.6) and after some developments we obtain

$$\ddot{e}_2(t) - \dot{e}_2(t)\left(2\lambda + 3\frac{\dot{e}_2(t)}{e_2(t)}\right) = 0 \tag{D.8}$$

which is a second order Liouville differential equation with two symmetric solutions

$$e_2(t) = \pm\frac{\sqrt{\lambda}}{\sqrt{-(C_1\exp^{2\lambda t} + 2C_2\lambda)}} \tag{D.9}$$

where C_1 and C_2 are integration constants. By plugging (D.9) into (D.7) the solutions for $e_3(t)$ are directly obtained

$$e_3(t) = \text{sgn}(e_2(t))\frac{\sqrt{\lambda(8C_2 - 1)}}{\sqrt{-(C_1\exp^{2\lambda t} + 2\lambda C_2}} \tag{D.10}$$

where $\text{sgn}(x)$ returns the sign of the given value. Then, by evaluating any of the two pairs of solutions at time $t = 0$, C_1 and C_2 can be expressed in terms of the initial conditions $e_2(0)$ and $e_3(0)$ leading finally to

$$e_2(t) = \frac{2e_2(0)}{a(t)} \tag{D.11}$$

$$e_3(t) = \frac{2e_3(0)}{a(t)} \tag{D.12}$$

with

$$a(t) = \sqrt{\left(e_2^2(0) + e_3^2(0)\right)\left(\exp^{2\lambda t} - 1\right) + 4\exp^{2\lambda t}} \tag{D.13}$$

Finally, let us plug (D.11) in the definition of $\dot{e}_1(t)$ in (D.2) and solving a first order differential equation with non-constant coefficients the following solution arises

$$e_1(t) = \frac{2e_1(0)}{a(t)} - \frac{2bZ^* \arctan\left(\frac{b(a(t)-2)}{b^2+2a(t)}\right)}{a(t)L} \tag{D.14}$$

with $b = \sqrt{e_2^2(0) + e_3^2(0)}$.

We note that function $a(t)$ has the following properties

$$\begin{aligned} a(0) &= 2 \\ \lim_{t\to\infty} a(t) &= \infty \end{aligned}$$

Furthermore, by looking at its derivative

$$\dot{a}(t) = \frac{\lambda \exp(2\lambda t)\left(e_2(0)^2 + e_3(0)^2 + 4\right)}{a(t)} \tag{D.15}$$

it is always positive. Therefore $a(t)$ is monotonic in $t \in [0,\,\infty)$ and is bounded in the interval $[2,\,\infty)$.

D.2 Study of the behavior of the depth vs. time

We now study the behavior of the depth control which depends on $e_1(t)$. We are interested on identifying under which analytic conditions it becomes a monotonic function. In order to achieve it, we are going to identify the extrema of $e_1(t)$ by studying when the first derivative zeroes. Then, we will search for sufficient conditions which ensure that the derivative never zeroes so that $e_1(t)$ is monotonic.

We remember that the expression of $e_1(t)$ is given by

$$e_1(t) = \frac{2e_1(0)}{a(t)} - \frac{2bZ^* \arctan\left(\frac{b(a(t)-2)}{b^2+2a(t)}\right)}{a(t)L} \tag{D.16}$$

whose derivative

$$\dot{e}_1(t) = -\frac{\lambda}{4L}\left(e_1(t)\left(4L + e_2(t)^2 L + e_3(t)^2 L\right) + 2Z^*\left(e_2(t)^2 + e_3(t)^2\right)\right) \tag{D.17}$$

From this expression it is evident that when $b = 0$ ($e_2(0) = 0$ and $e_3(0) = 0$ so that the camera is already parallel to the object) a linear differential equation is obtained so that $e_1(t)$ is monotonic. Otherwise, it is necessary to study the derivative of $e_1(t)$ which can be rewritten as

$$\dot{e}_1(t) = -\frac{-2\dot{a}(t)\left(-bZ^* \arctan\left(\frac{b(a(t)-2)}{b^2+2a(t)}\right)\left(b^2+a(t)^2\right) + e_1(0)L\left(b^2+a(t)^2\right) + b^2 Z^* a(t)\right)}{L\left(b^2+a(t)^2\right)a(t)^2}$$

(D.18)

As shown in Appendix D.1, $\dot{a}(t)$ is always positive and never zeroes, while the denominator of $\dot{e}_1(t)$ is also positive. Therefore, $\dot{e}_1(t)$ only zeroes when

$$-bZ^* \arctan\left(\frac{b(a(t)-2)}{b^2+2a(t)}\right)\left(b^2+a(t)^2\right) + e_1(0)L\left(b^2+a(t)^2\right) + b^2 Z^* a(t) = 0 \quad \text{(D.19)}$$

By setting the following change of variable

$$u(t) = \frac{b(a(t)-2)}{b^2+2a(t)}$$

(D.20)

the expression in (D.19) can be rewritten as

$$\arctan(u(t)) = f(u(t))$$

(D.21)

with

$$f(u(t)) = \frac{u(t)^2\left(e_1(0)L\left(b^2+4\right) - 2b^2 Z^*\right) + u(t)bZ^*\left(b^2-4\right)e_1(0)L\left(b^2+4\right) + 2b^2 Z^*}{bZ^*\left((b^2+4)(u(t)^2+1)\right)}$$

(D.22)

Note that the derivative of $e_1(t)$ only zeroes if and only if $\arctan(u(t))$ intersects with $f(u(t))$. Therefore, if we can find analytical conditions which avoid both functions to intersect, $e_1(t)$ will be monotonic under those conditions since its derivative never will zero.

We first study the behavior of $u(t)$. The following properties hold

$$\dot{u}(t) = \dot{a}(t)\frac{b\left(b^2+4\right)}{\left(b^2+2a(t)\right)^2}$$

(D.23)

$$u(0) = 0$$

(D.24)

$$\lim_{t\to\infty} u(t) = b/2$$

(D.25)

and since $a(t)$ is strictly monotonic increasing then $u(t)$ is also strictly monotonic increasing when $b \neq 0$ and it is bounded in the interval $[0, b/2]$ for $t \in [0, \infty)$.

The derivative of $f(u(t))$ is

$$\dot{f}(u(t)) = -\dot{u}(t)\frac{u(t)^2\left(b^2-4\right) + u(t)8b + 4 - b^2}{(u(t)+1)\left((b^2+4)(u(t)^2+1)\right)}$$

(D.26)

since $\dot{u}(t)$ is always positive and so is the denominator of $\dot{f}(u(t))$, the sign of $\dot{f}(u(t))$ depends on the following polynomial

$$p(u(t)) = u(t)^2(b^2 - 4) + u(t)8b + 4 - b^2 \tag{D.27}$$

which can be written as

$$\left(b^2 - 4\right)(u - u_1)(u - u_2) \tag{D.28}$$

with u_1 and u_2 the roots of the polynomial

$$
\begin{aligned}
u_1 &= -\frac{b+2}{b-2} \\
u_2 &= \frac{b-2}{b+2}
\end{aligned}
\tag{D.29}
$$

When $p(u(t)) < 0$ then $f(u(t))$ increases and inversely. Note that depending on if $b < 2$ or $b > 2$ the sign of $p(u(t))$ is affected. Let us study the sign depending on these conditions.

b¡2: in this case $b^2 - 4$ is negative, and $u_1 > b/2$ and $u_2 < 0$ as shown hereafter

$$u_1 > \frac{1}{2} \Leftrightarrow \frac{1}{2} - u_1 < 0 \equiv \frac{1}{2} + \frac{b+2}{b-2} < 0 \equiv \frac{b^2+4}{2(b-2)} < 0 \tag{D.30}$$

$$u_2 < 0 \Leftrightarrow \frac{b-2}{b+2} < 0 \tag{D.31}$$

which means that when $b < 2$ there are no zero-crossings in the interval $u \in [0, b/2]$. Furthermore, the sign of the polynomial in this interval is always

$$\underbrace{\left(b^2 - 4\right)}_{<0}\underbrace{(u - u_1)}_{<0}\underbrace{(u - u_2)}_{>0} > 0 \tag{D.32}$$

b¿2: in this case $b^2 - 4$ is positive, and $u_1 < 0$ and $u_2 \in [0, b/2]$ since

$$u_1 < 0 \Leftrightarrow -\frac{b+2}{b+2} < 0 \tag{D.33}$$

$$
\begin{cases}
u_2 > 0 \Leftrightarrow \dfrac{b-2}{b+2} > 0 \\[2mm]
u_2 < \dfrac{b}{2} \Leftrightarrow \dfrac{b}{2} - u_2 > 0 \equiv \dfrac{b^2+4}{2(b+2)} > 0
\end{cases}
\tag{D.34}
$$

Therefore $p(u(t))$ has a unique zero-crossing in the interval $u \in [0, b/2]$ so that $p(u(t))$ is always increasing in such interval because

$$
\begin{aligned}
p(u = 0) \;=&\; \underbrace{(b^2 - 4)}_{>0}\,\underbrace{(u_1)}_{<0}\,\underbrace{(u_2)}_{>0} < 0 \\
p(u = b/2) \;=&\; \underbrace{(b^2 - 4)}_{>0}\,\underbrace{(b/2 - u_1)}_{>0}\,\underbrace{(b/2 - u_2)}_{>0} > 0
\end{aligned}
\tag{D.35}
$$

Then, the behavior of $f(u(t))$ is determined by the following two cases (we remember that $b \geq 0$ and when $b = 0$ $e_1(t)$ is always monotonic)

- $b \in (0, 2]$: $f(u(t))$ is monotonic decreasing when $u(t) \in [0, b/2]$.

- $b > 2$: $f(u(t))$ has a global maximum in the interval $[0, b/2]$ when $u(t) = (b-2)/(b+2)$.

In summary, the behavior of $\arctan(u(t))$ and $f(u(t))$ in the interval $[0, b/2]$ is represented in Fig. (D.1).

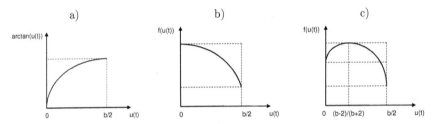

Figure D.1: Schema of $\arctan(u(t))$ and $f(u(t))$ when $u(t) \in [0, b/2]$. a) $\arctan(u(t))$. b) $f(u(t))$ when $b \in (0, 2]$. c) $f(u(t))$ when $b > 2$.

Note that two sufficient conditions can be defined in order to avoid the intersection of $\arctan(u(t))$ and $f(u(t))$:

- $\min(f(u(t))) > \max(\arctan(u(t)))$
- $\max(f(u(t))) < \min(\arctan(u(t)))$

According to the first condition we need the following expressions

$$
\min(f(u(t))) \;=\; f(u = b/2) = \frac{e_1(0)L}{b\arctan(b/2)}
\tag{D.36}
$$

$$
\max(\arctan(u(t))) \;=\; \arctan(b/2)
\tag{D.37}
$$

so that the condition is

$$
Z^* < -\frac{e_1(0)L}{b\arctan(b/2)}
\tag{D.38}
$$

The second condition requires to distinguish between two cases

- $b \in (0, 2]$: $\max(f(u(t))) = f(u = 0) = \dfrac{e_1(0)L(b^2 + 4) + 2b^2 Z^*}{bZ^*(b^2 + 4)}$

- $b > 2$: $\max(f(u(t))) = f\left(u = \frac{b-2}{b+2}\right) = \dfrac{bZ^* + 2Le_1(0)}{2bZ^*}$

while

$$\min(\arctan(u(t))) = 0 \qquad \text{(D.39)}$$

Therefore, the second sufficient condition is

$$
b \in (0, 2] \quad \rightarrow \quad Z^* < -\frac{e_1(0)L(b^2 + 4)}{2b^2}
$$
$$
b > 2 \quad \rightarrow \quad Z^* < -\frac{2Le_1(0)}{b} \qquad \text{(D.40)}
$$

If these sufficient conditions are true then no intersection between $\arctan(u(t))$ and $f(u(t))$ will occur, and therefore, $e_1(t)$ will be monotonic.

The sufficient conditions in (D.38) and in (D.40) are expressed in terms of the initial state of the task function. They can be rewritten in terms of the initial object pose in the camera frame obtaining two different cases. The first case corresponds to the condition in (D.38) which is valid when the camera must go forward ($Z^* < -D(0)$). In such a case, the sufficient condition to ensure that $e_1(t)$ is monotonic is

$$
Z^* < \frac{-2D(0)}{C(0)\left(2 + \sqrt{1 - C(0)^2} \arctan\left(\frac{\sqrt{1-C(0)^2}}{2}\right)\right)} \qquad \text{(D.41)}
$$

On the other hand, from (D.40) we obtain the sufficient condition valid when the camera must go backwards since $Z^* > -D(0)$

$$
b \in (0, 2] \quad \rightarrow \quad Z^* > -\frac{D(0)}{C(0)^3}
$$
$$
b > 2 \quad \rightarrow \quad Z^* > \frac{-2D(0)}{2C(0) - \sqrt{1 - C(0)^2}} \qquad \text{(D.42)}
$$

In summary, we have obtained sufficient conditions depending on the initial state (or the initial camera-object pose) which ensure that $e_1(t)$ will be monotonic. Even if these sufficient conditions are not ensured, we can al least ensure that $e_1(t)$ will present a unique peak since a unique intersection of $\arctan(u(t))$ and $f(u(t))$ occurs.

Bibliography

Agapakis, J. E. (1990). Approaches for recognition and interpretation of workpiece surface features using structured lighting. *Int. Journal of Robotics Research* **9**(5), 3–16.

Albus, J., E. Kent, M. Nashman, P. Manabach and L. Palombo (1982). Six-dimensional vision system. In: *Proc. SPIE.* Vol. 336 of *Robot Vision.* pp. 142–153.

Alhaj, A., C. Collewet and F. Chaumette (2003). Visual servoing based on dynamic vision. In: *IEEE Int. Conf. on Robotics and Automation.* Taipei, Taiwan. pp. 3055–3060.

Aloimonos, Y., I. Weiss and A. Bandopadhay (1987). Active vision. *Int. Journal of Computer Vision* **1**(4), 333–356.

Amin-Nejad, S., J. S. Smith and J. Lucas (2003). A visual servoing system for edge trimming of fabric embroideries by laser. *Mechatronics* **13**(6), 533–551.

Andreff, N., B. Espiau and R. Horaud (2002). Visual servoing from lines. *Int. Journal of Robotics Research* **21**(8), 679–700.

Armangué, X. and J. Salvi (2003). Overall view regarding funtamental matrix estimation. *Image and Vision Computing* **21**(2), 205–220.

Armangué, X., J. Salvi and H. Araújo (2003). A review on egomotion by means of differential epipolar geomety applied to the movement of a mobile robot. *Pattern Recognition* **21**(12), 2927–2944.

Audin, M. (2002). *Geometry.* Springer.

Basri, R., E. Rivlin and I. Shimshoni (1998). Visual homing: surfing on the epipoles. In: *IEEE Int. Conf. on Computer Vision.* Bombay, India. pp. 863–869.

Batista, J., H. Araújo and A. T. Almeida (1999). Iterative multistep explicit camera calibration. *IEEE Trans. on Robotics and Automation* **15**(5), 897–916.

Batlle, J., E. Mouaddib and J. Salvi (1998). Recent progress in coded structured light as a technique to solve the correspondence problem: a survey. *Pattern Recognition* **31**(7), 963–982.

Benhimane, S. and E. Malis (2003). Vision-based control with respect to planar and non-planar objects using a zooming camera. In: *Int. Conf. on Advanced Robotics*. Coimbra, Portugal. pp. 991–996.

Bergmann, D. (1995). New approach for automatic surface reconstruction with coded light. In: *Remote Sensing and Reconstruction for Three-Dimensional Objects and Scenes* (T. F. Schenck, Ed.). Vol. 2572. San Diego. pp. 2–9.

Besl, Paul J. (1988). *Advances in Machine Vision: Architectures and Applications*. Chap. Active Optical Range Imaging Sensors, pp. 1–53. Springer-Verlag.

Bien, Z., W. Jang and J. Park (1993). *Visual servoing*. Chap. Characterization and use of feature-jacobian matrix for visual servoing, pp. 317–363. World Scientific.

Blais, F. and M. Rioux (1986). Real-time numerical peak detector. *Signal Procesing* **11**, 145–155.

Boyer, K. L. and A. C. Kak (1987). Color-encoded structured light for rapid active ranging. *IEEE Trans. on Pattern Analysis and Machine Intelligence* **9**(1), 14–28.

Buendía, M., R. Salvador, R. Cibrián, M. Laguia and J. M. Sotoca (1999). Determination of the object surface function by structured light: application to the study of spinal deformities. *Physics in Medicine and Biology* **44**(1), 75–86.

Capek, K. (1970). *R.U.R. (Rossum's Universal Robots)*. Pocket Books.

Carrihill, B. and R. Hummel (1985). Experiments with the intensity ratio depth sensor. *Computer Vision, Graphics and Image Processing* **32**, 337–358.

Caspi, D., N. Kiryati and J. Shamir (1998). Range imaging with adaptive color structured light. *IEEE Trans. on Pattern Analysis and Machine Intelligence* **20**(5), 470–480.

Chaumette, F. (1998). Potential problems of stability and convergence in image-based and position-based visual servoing. In: *The Conference of Vision and Control*. Vol. 237. LNCIS Series. pp. 66–78.

Chaumette, F. (2002). *La commande des robots manipulateurs*. Chap. 3 - Asservissement visuel, pp. 105–150. Traité IC2. W. Khalil (ed.), Hermès.

Chaumette, F. (2004). Image moments: a general and useful set of features for visual servoing. *IEEE Trans. on Robotics* **20**(4), 713–723.

Chazan, G. and N. Kiryati (1995). Pyramidal intensity-ratio depth sensor. Technical report 121. Center for Communication and Information Technologies. Department of Electrical Engineering, Technion, Haifa, Israel.

Chen, C., Y. Hung, C. Chiang and J. Wu (1997). Range data acquisition using color structured lighting and stereo vision. *Image and Vision Computing* **15**, 445–456.

Chen, F., G.M. Brown and M. Song (2000). Overview of three-dimensional shape measurement using optical methods. *Optical Engineering* **1**(39), 10–22.

Chesi, G., E. Malis and R. Cipolla (2000). Automatic segmentation and matching of planar contours for visual servoing. In: *IEEE Int. Conf. on Robotics and Automation.* Vol. 3. S. Francisco, CA, United States. pp. 2753–2758.

Cipolla, R. and E. Blake (1992). Surface shape from the deformation of apparent contours. *Int. Journal of Computer Vision* **9**(2), 83–112.

Clocksin, W. F., J. S. E Bromley, P. G. Davey, A. R. Vidler and Morgan C. G. (1985). An implementation of model-based visual feedback for robot arc welding of thin sheet metal. *Int. Journal of Robotics Research* **4**(1), 13–26.

Collewet, C., A. Alhaj and F. Chaumette (2004). Model-free visual servoing on complex images based on 3d reconstruction. In: *IEEE Int. Conf. on Robotics and Automation.* Vol. 1. pp. 751–756.

Collewet, C. and F. Chaumette (2000). A contour approach for image-based control of objects with complex shape. In: *IEEE/RSJ Int. Conf. on Intelligent Robots and Systems.* Vol. 1. Takamatsu, Japan. pp. 751–756.

Collewet, C. and F. Chaumette (2002). Positioning a camera with respect to planar objects of unknown shape by coupling 2-d visual servoing and 3-d estimations. *IEEE Trans. on Robotics and Automation* **3**(18), 322–333.

Colombo, C. and B. Allotta (1999). Image-based robot task planning and control using a compact visual representation. *IEEE Trans. on Systems, Man, and Cybernetics* **29**(1), 92–100.

Colombo, C., B. Allota and P. Dario (1995). Affine visual servoing: a framework for relative positionning with a robot. In: *IEEE Int. Conf. on Robotics and Automation.* Vol. 1. Nagoya, Japan. pp. 21–27.

Corke, P. I. and S. A. Hutchinson (2001). A new partitioned approach to image-based visual servo control. *IEEE Trans. on Robotics and Automation* **17**(4), 507–515.

Cox, I. J., S. L. Hingorani, S. B. Rao and B. M. Maggs (1996). A maximum likelihood stereo algorithm. *Computer Vision and Image Understanding* **63**(3), 542–567.

Crétual, A. and F. Chaumette (2001). Visual servoing based on image motion. *Int. Journal of Robotics Research* **20**(11), 857–877.

Davies, C. J. and M. S. Nixon (1998). A hough transform for detecting the location and orientation of 3-dimensional surfaces via color encoded spots. *IEEE Trans. on systems, man and cybernetics* **28**(1), 90–95.

De la Escalera, A., L. Moreno, M. A. Salichs and J. M. Armingol (1996). Continuous mobile robot localizatin by using structured light and geometric map. *Int. Journal of Systems Science* **27**(8), 771–782.

De Ma, S. (1993). Conics-based stereo, motion estimation and pose determination. *Int. Journal of Computer Vision* **10**(1), 7–25.

Deguchi, K. (1997). Direct interpretation of dynamic images and camera motion for visual servoing without image feature correspondence. *Journal of Robotics and Mechatronics* **9**(2), 104–110.

Dementhon, D. and L. S. Davis (1995). Model-based object pose in 25 lines of code. *Int. Journal of Computer Vision* **15**(1/2), 123–141.

DeSouza, G. N. and A. C. Kak (2002). Vision for mobile robot navigation: a survey. *IEEE Trans. on Pattern Analysis and Machine Intelligence* **24**(2), 237–267.

Dhome, M., J. T. Lapresté, G. Rives and M. Richetin (1990). Spatial localization of modelled objects of revolution in monocular perspective vision. In: *European Conf. on Computer Vision*. Vol. 427. Antibes. pp. 475–485.

Dhome, M., M. Richetin, J.T. LaPresté and G. Rives (1989). Determination of the attitude of 3-d objects from a single perspective view. *IEEE Trans. on Pattern Analysis and Machine Intelligence* **11**(12), 1265–1278.

Drummond, T. W. and R. Cipolla (1999). Visual tracking and control using lie algebras. In: *IEEE Int. Conf. on Computer Vision and Pattern Recognition*. Vol. 2. Fort Collins, Colorado, USA. pp. 652–657.

Drummond, T. W. and R. Cipolla (2002). Real-time tracking of complex structures with on-line camera calibration. *Image and Vision Computing* **20**(5-6), 427–433.

Dubrawski, A. and I. Siemiatkowska (1998). A method for tracking the pose of a mobile robot equipped with a scanning laser range finder. In: *IEEE Int. Conf. on Robotics and Automation*. Vol. 3. Leuven, Belgium. pp. 2518–2523.

Durdle, N. G., J. Thayyoor and V. J. Raso (1998). An improved structured light technique for surface reconstruction of the human trunk. In: *IEEE Canadian Conf. on Electrical and Computer Engineering*. Vol. 2. pp. 874–877.

Espiau, B., F. Chaumette and P. Rives (1992). A new approach to visual servoing in robotics. *IEEE Trans. on Robotics and Automation* **8**(6), 313–326.

Etzion, T. (1988). Constructions for perfect maps and pseudorandom arrays. *IEEE Transactions on information theory* **34**(5), 1308–1316.

Faugeras, O. (1993). *Three-Dimensional Computer Vision*. MIT Press. Massachusetts Institute of Technology.

Feddema, J. T., C. S. G. Lee and O. R. Mitchell (1991). Weighted selection of image features for resolved rate visual feedback control. *IEEE Trans. on Robotics and Automation* **7**(1), 31–47.

Fofi, D., T. Sliwa and Y. Voisin (2004). A comparative survey on invisible structured light. In: *Machine vision applications in industrial inspection XII.* Vol. 5303. San José, USA. pp. 90–98.

Forest, J. and J. Salvi (2002). An overview of laser slit 3d digitasers. In: *IEEE/RSJ Int. Conference on Robots and Systems.* Lausanne, Switzerland.

Fredricksen, H. (1982). A survey of full length nonlinear shift register cycle algorithms. *Society of Industrial and Applied Mathematics Review* **24**(2), 195–221.

Fusiello, A., E. Trucco and A. Verri (2000). A compact algorithm for rectification of stereo pairs. *Machine Vision and Applications* **12**(1), 16–22.

Geng, Z. J. (1996). Rainbow 3-dimensional camera: New concept of high-speed 3-dimensional vision systems. *Optical Engineering* **35**(2), 376–383.

Griffin, P.M., L.S. Narasimhan and S.R. Yee (1992). Generation of uniquely encoded light patterns for range data acquisition. *Pattern Recognition* **25**(6), 609–616.

Grimson, W. E. L., T. Lozano-Perez, N. Noble and S.J. White (1993). An automatic tube inspection system that finds cylinders in range data. In: *IEEE Int. Conf. on Computer Vision and Pattern Recognition.* Vol. 1. New York, NY, USA. pp. 446–452.

Gühring, Jens (2001). Dense 3-d surface acquisition by structured light using off-the-shelf components. In: *Photonics West, Videometrics VII.* Vol. 4309. pp. 220–231.

Gutsche, R., T. Stahs and F. M. Wahl (1991). Path generation with a universal 3d sensor. In: *IEEE Int. Conf. on Robotics and Automation.* Vol. 1. Sacramento, CA, USA. pp. 838–843.

Hager, G. D. (1997). A modular system for robust positioning using feedback from stereo vision. *IEEE Trans. on Robotics and Automation* **13**(4), 582–595.

Hall-Holt, O. and S. Rusinkiewicz (2001). Stripe boundary codes for real-time structured-light range scanning of moving objects. In: *IEEE Int. Conference on Computer Vision.* pp. II: 359–366.

Haralick, R.M., H. Joo, C.N. Lee, X. Zhuang, V.G. Vaidya and M.B. Kim (1989). Pose estimation from corresponding point data. *IEEE Trans. on Systems, Man and Cybernetics* **19**(6), 1426–1446.

Hartley, R. and A. Zisserman (2000). *Multiple View Geometry in Computer Vision.* Cambridge University Press.

Haverinen, J. and J.; Roning (1998). An obstacle detection system using a light stripe identification based method. In: *IEEE Int. Joint Symp. on Intelligence and Systems.* Vol. 1. Rockville, MD, USA. pp. 232–236.

Horaud, R., B. Conio, O. Leboulleux and B. Lacolle (1989). An analytic solution for the perspective 4-point problem. *Computer Vision, Graphics, and Image processing* **47**(1), 33–44.

Horn, E. and N. Kiryati (1999). Toward optimal structured light patterns. *Image and Vision Computing* **17**(2), 87–97.

Hsieh, Y. C. (2001). Decoding structured light patterns for three-dimensional imaging systems. *Pattern Recognition* **34**, 343–349.

Huang, T. S. and A. N. Netravali (1994). Motion and structure from feature correspondences: a review. *Proceedings of the IEEE* **82**(2), 252–268.

Hügli, H. and G. Maître (1989). Generation and use of color pseudo random sequences for coding structured light in active ranging. In: *Industrial Inspection.* Vol. 1010. pp. 75–82.

Hung, D.C.D. (1993). 3d scene modelling by sinusoid encoded illumination. *Image and Vision Computing* **11**, 251–256.

Hutchinson, S., G. Hager and P.I. Corke (1996). A tutorial on visual servo control. *IEEE Trans. on Robotics and Automation* **12**(5), 651–670.

Inokuchi, S., K. Sato and F. Matsuda (1984). Range imaging system for 3-D object recognition. In: *IEEE Int. Conf. on Pattern Recognition.* pp. 806–808.

Ito, M. and A. Ishii (1995). A three-level checkerboard pattern (tcp) projection method for curved surface measurement. *Pattern Recognition* **28**(1), 27–40.

Jarvis, R. (1993). Range sensing for computer vision. *Advances in Image Communications. Elsevier Science Publishers, Amsterdam* pp. 17–56.

Joung, I. S. and H. S. Cho (1998). An active omni-directional range sensor for mobile robot navigation. *Control Engineering Practice* **6**(3), 385–393.

Jung, M. J., H. Myung, S. G. Hong, D. R. Park, H. K. Lee and S. Bang (2004). Structured light 2d range finder for simultaneous localization and map-building (slam) in home environments. In: *IEEE Int. Symposium on Micro-Nanomechatronics and Human Science.* Vol. 1. Nagoya, Japan. pp. 371–376.

Kahane, B. and Y. Rosenfeld (2004). Real-time 'sense-and-act' operation for construction robots. *Automation in Construction* **13**(6), 751–764.

Kemmotsu, K. and T. Kanade (1995). Uncertainty in object pose determination with three light-stripe range measurements. *IEEE Trans. on Robotics and Automation* **11**(5), 741–747.

Khadraoui, D., G. Motyl, P. Martinet, J. Gallice and F. Chaumette (1996). Visual servoing in robotics scheme using a camera/laser-stripe sensor. *IEEE Trans. on Robotics and Automation* **12**(5), 743–750.

Kim, K. H. and H. S. Cho (2001). Range and contour fused environment recognition for mobile robot. In: *IEEE Int. Conf. on Multisensor Fusion and Integration for Intelligent Systems*. Vol. 1. Baden-Baden, Germany. pp. 183–188.

Kim, P., S. Rhee and C. H. Lee (1999). Automatic teaching of welding robot for free-formed seam using laser vision sensor. *Optics and Lasers in Engineering* **31**(3), 173–182.

Kiyasu, S., H. Hoshino, K. Yano and S. Fujimura (1995). Measurement of the 3-D shape of specular polyhedrons using an m-array coded light source. *IEEE Trans. on Instrumentation and Measurement* **44**(3), 775–778.

Kondo, H. and U. Tamaki (2004). Navigation of an auv for investigation of underwater structures. *Control engineering practice* **12**(12), 1551–1559.

Krupa, A., J. Gangloff, C. Doignon, M. Mathelin, G. Morel, J. Leroy, L. Soler and J. Marescaux (2003). Autonomous 3d positionning of surgical instruments in robotized laparoscopic surgery using visual servoing. *IEEE Trans. on Robotics and Automation* **19**(5), 842–853.

Kwok, K. S., C. S. Loucks and B.J. Driessen (1998). Rapid 3-d digitizing and tool path generation for complex shapes. In: *IEEE Int. Conf. on Robotics and Automation*. Vol. 4. Leuven, Belgium. pp. 2789–2794.

Lavoie, P., D. Ionescu and E. Petriu (1999). A high precision 3D object reconstruction method using a color coded grid and nurbs. In: *Int. Conference on Image Analysis and Processing*. Venice, Italy. pp. 370–375.

Le Moigne, J. J. and A. M. Waxman (1988). Structured light patterns for robot mobility. *IEEE Trans. on Robotics and Automation* **4**(5), 541–548.

Levoy, M., K. Pulli, B. Curless, S. Rusinkiewicz, D. Koller, L. Pereira, M. Ginzton, S. E. Anderson, J. Davis, J. Ginsberg, J. Shade and D. Fulk (2000). The digital michelangelo project: 3d scanning of large statues. In: *Int. Conf. on Computer Graphics and Interactive Techniques*. Vol. 1. New Orleans, Louisiana, USA. pp. 131–144.

Lin, X., J. Zeng and Q. Yao (1996). Optimal sensor planning with minimal cost for 3d object recognition using sparse structured light images. In: *IEEE Int. Conf. on Robotics and Automation*. Vol. 4. Minneapolis, MN, USA. pp. 3484–3489.

Lots, J.F., D.M. Lane and E. Trucco (2000). Application of 2 1/2 d visual servoing to underwater vehicle station-keeping. In: *MTS/IEEE OCEANS Conf.*. Vol. 2. Providence, USA. pp. 1257–1264.

Lots, J.F., D.M. Lane, E. Trucco and F. Chaumette (2001). A 2d visual servoing for underwater vehicle station keeping. In: *IEEE int. Conf. on Robotics and Automation*. Vol. 3. Seoul, Korea. pp. 2767–2772.

Lowe, D. G. (1991). Fitting parameterized three-dimensional models to images. *IEEE Trans. on Pattern Analysis and Machine Intelligence* **13**(5), 441–450.

Lozano-Perez, T., J. Jones, E. Mazer, P. O'Donnell, W. Grimson, P. Tournassoud and A. Lanusse (1987). Handey: A robot system that recognizes, plans, and manipulates. In: *IEEE Int. Conf. on Robotics and Automation*. Vol. 4. Raleigh, North Carolina, USA. pp. 843–849.

MacWilliams, F. J. and N. J. A. Sloane (1976). Pseudorandom sequences and arrays. *Proceedings of the IEEE* **64**(12), 1715–1729.

Mahony, R., P. Corke and F. Chaumette (2002). Choice of image features for depth-axis control in image-based visual servo control. In: *IEEE/RSJ Int. Conf. on Intellingent Robots and Systems*. Lausanne, Switzerland.

Malis, E. and F. Chaumette (2000). 2 1/2 d visual servoing with respect to unknown objects through a new estimation scheme of camera displacement. *Int. Journal of Computer Vision* **37**(1), 79–97.

Malis, E. and F. Chaumette (2002). Theoretical improvements in the stability analysis of a new class of model-free visual servoing methods. *IEEE Trans. on Robotics and Automation* **18**(2), 176–186.

Malis, E. and P. Rives (2003). Robustness of image-based visual servoing with respect to depth distribution errors. In: *IEEE Int. Conf. on Robotics and Automation*. Vol. 1. Taipei, Taiwan. pp. 1056–1061.

Malis, E., F. Chaumette and S. Boudet (1999). 2 1/2 d visual servoing. *IEEE Trans. on Robotics and Automation* **15**(2), 238–250.

Malis, E., J.J. Borrely and P. Rives (2002). Intrinsics-free visual servoing with respect to straight lines. In: *IEEE/RSJ Int. Conf. on Intelligent Robots and Systems*. Vol. 1. Laussane, Switzerland. pp. 384–389.

Marchand, E. and F. Chaumette (1999). Active vision for complete scene reconstruction and exploration. *IEEE Trans. on Pattern Analysis and Machine Intelligence* **21**(1), 65–72.

Maruyama, M. and S. Abe (1993). Range sensing by projecting multiple slits with random cuts. *IEEE Trans. on Pattern Analysis and Machine Intelligence* **15**(6), 647–651.

Matthies, L., Y. Xionga, R. Hogga, D. Zhua, A. Rankina, B. Kennedya, M. Hebertb, R. Maclachlanb, C. Wonc, T. Frostc, G. Sukhatmed, M. McHenryd and S. Goldberg (2002). A portable, autonomous, urban reconnaissance robot. *Robotics and Autonomous Systems* **40**(2–3), 163–172.

Mezouar, Y. and F. Chaumette (2002). Path planning for robust image-based control. *IEEE Trans. on Robotics and Automation* **18**(4), 534–549.

Minou, M., T. Kanade and T. Sakai (1981). A method of time-coded parallel planes of light for depth measurement. *Transactions of the IECE of Japan* **64**, 521–528.

Miyasaka, T., K. Kuroda, M. Hirose and K. Araki (2000). High speed 3-D measurement system using incoherent light source for human performance analysis. In: *Congress of The International Society for Photogrammetry and Remote Sensing*. The Netherlands, Amsterdam. pp. 65–69.

Monks, T. P., J. N. Carter and C. H. Shadle (1992). Colour-encoded structured light for digitisation of real-time 3D data. In: *IEEE Int. Conf. on Image Processing*. pp. 327–30.

Monks, T.P. and J.N Carter (1993). Improved stripe matching for colour encoded structured light. In: *Int. Conf. on Computer Analysis of Images and Patterns*. pp. 476–485.

Morano, R. A., C. Ozturk, R. Conn, S. Dubin, S. Zietz and J. Nissanov (1998). Structured light using pseudorandom codes. *IEEE Trans. on Pattern Analysis and Machine Intelligence* **20**(3), 322–327.

Morel, G., T. Leibezeit, J. Szewczyk, S. Boudet and J. Pot (2000). Explicit incorporation of 2d constraints in vision-based control of robot manipulators. In: *Int. Symp. on Experimental Robotics*. Vol. 250 of *LNCIS*. Springer Verlag. Sidney, Australia. pp. 99–108.

Morita, H., K. Yajima and S. Sakata (1988). Reconstruction of surfaces of 3-d objects by m-array pattern projection method. In: *IEEE Int. Conference on Computer Vision*. pp. 468–473.

Motyl, G. (1992). Couplage d'une caméra et d'un faisceau laser en commande référencée vision. PhD thesis. Thèse de l'Université Blaise Pascale Clermont-Ferrand.

Nayar, S. K., H. Peri, M. Grossber and P. Belhumeur (2003). A projector with automatic radiometric screen compensation. Technical report. Department of Computer Science, Columbia University. New York.

Neira, J., J. D. Tardos, J. Horn and G. Schmidt (1999). Fusing range and intensity images for mobile robot localization. *IEEE Trans. on Robotics and Automation* **15**(1), 76–84.

Nevado, M. M., J. Gómez and E. Zalama (2004). Obtaining 3d models of indoor environments with a mobile robot by estimating local surface directions. *Robotics and Autonomous Systems* **48**(2-3), 131–143.

Niel, A., P. Sommer, S. H. Kolpl and Y. Lypetskyy (2004). High precision measurement of free formed parts using industrial robots. In: *IEEE Int. Workshop on Robot Sensing*. Graz, Austria. pp. 79–84.

Nurre, J. H., E. L. Hall and J. J. Roning (1988). Acquiring simple patterns for surface inspection. In: *IEEE Int. Conf. on Computer Vision and Pattern Recognition*. Vol. 1. Ann Arbor, MI, USA. pp. 586–591.

Nwodoh, T., B. Nnaji, R. Popplestone and E. Lach (1997). Three-dimensional model acquisition for medical robotics-assisted burn debridement system. *Robotics and Computer-Integrated Manufacturing* **13**(4), 309–318.

Pagès, J., C. Collewet, F. Chaumette and J. Salvi (2004). Plane-to-plane positioning from image-based visual servoing and structured light. In: *IEEE/RSJ Int. Conf. on Intelligent Robots and Systems*. Vol. 1. Sendai, Japan. pp. 1004–1009.

Pajdla, T. (1995). Bcrf - binary-coded illumination range finder reimplementation. Technical report KUL/ESAT/MI2/9502. Katholieke Universiteit Leuven. ESAT, Kardinaal Mercierlaan 94, B-3001 Leuven.

Perez de la Blanca, N., J.M. Fuertes and M.J Lucena (2003). 3d rigid facial motion estimation from disparity maps. In: *Iberoamerican Congress on Pattern Recognition*. Vol. LNCS 2905. Havana, Cuba. pp. 54–61.

Petriu, E. M., T. Bieseman, N. Trif, W. S. McMath and S. K. Yeung (1992). Visual object recognition using pseudo-random grid encoding. In: *IEEE/RSJ Int. Conf. on Intelligent Robots and Systems*. pp. 1617–1624.

Petriu, E. M., Z. Sakr, Spoelder H. J. W. and A. Moica (2000). Object recognition using pseudo-random color encoded structured light. In: *IEEE Instrumentation and Measurement technology Conference*. Vol. 3. pp. 1237–1241.

Posdamer, J. L. and M. D. Altschuler (1982). Surface measurement by space-encoded projected beam systems. *Computer Vision, Graphics and Image Processing* **18**(1), 1–17.

Questa, P., E. Grossmann and G. Sandini (1995). Camera self orientation and docking maneuver using normal flow. In: *SPIE AeroSense*. Orlando, Florida.

Ramachandram, D. and M. Rajeswari (2000). Structured lighting to enhance global image feature sensitivity in a neural network based robot-positioning task. In: *Intelligent Systems and Technologies for the New Millennium*. Vol. 2. Kuala Lumpur, Malaysia. pp. 80–85.

Rives, P. (2000). Visual servoing based on epipolar geometry. In: *IEEE/RSJ Int. Conf. on Intelligent Robots and Systems*. Vol. 1. Takamatsu, Japan. pp. 602–607.

Rocchini, C., Paulo Cignoni, C. Montani, P. Pingi and Roberto Scopigno (2001). A low cost 3D scanner based on structured light. In: *EuroGraphics* (A. Chalmers and T.-M. Rhyne, Eds.). Vol. 20(3). pp. 299–308. Blackwell Publishing.

Rutkowski, W., R. Benton and E. Kent (1987). Model-driven determination of object pose for a visually servoed robot. In: *IEEE Int. Conf. on Robotics and Automation*. Vol. 4. St. Louis, Missouri, USA. pp. 1419–1428.

Sagan, H. (1994). *Space Filling Curves*. Springer. New York.

Salvi, J., J. Batlle and E. Mouaddib (1998). A robust-coded pattern projection for dynamic 3d scene measurement. *Pattern Recognition Letters* **19**(11), 1055–1065.

Salvi, J., J. Pagès and J. Batlle (2004). Pattern codification strategies in structured light systems. *Pattern Recognition* **37**(4), 827–849.

Salvi, J., X. Armangué and J. Batlle (2002). A comparative review of camera calibrating methods with accuracy evaluation. *Pattern Recognition* **35**(7), 1617–1635.

Samson, C., M. Le Borgne and B. Espiau (1991). *Robot control: the task function approach.* Clarendon Press, Oxford.

Sansoni, G., S. Lazzari, S. Peli and F. Docchio (1997). 3d imager for dimensional gauging of industrial workpieces: state of the art of the development of a robust and versatile system. In: *Int. Conf. on Recent Advances in 3-D Digital Imaging and Modeling.* Otawwa, Ontario, Canada. pp. 19–26.

Santos-Victor, J. and G. Sandini (1997). Visual behaviors for docking. *Computer Vision and Image Understanding* **67**(3), 223–238.

Sato, K. (1996). Range imaging based on moving pattern light and spatio-temporal matched filter. In: *IEEE Int. Conf. on Image Processing.* Vol. 1. pp. 33–36.

Sato, T. (1999). Multispectral pattern projection range finder. In: *Conf. on Three-Dimensional Image Capture and Applications II.* Vol. 3640. San Jose, California. pp. 28–37.

Sazbona, D., Z. Zalevskyb and E. Rivlin (2005). Qualitative real-time range extraction for preplanned scene partitioning using laser beam coding. *Pattern Recognition Letters* **26**(11), 1772–1781.

Schilling, R. J. (1990). *Fundamentals of Robotics: analysis and control.* Prentice-Hall International.

Schramm, F., G. Morel, A. Micaelli and A. Lottin (2004). Extended-2d visual servoing. In: *IEEE Int. Conf. on Robotics and Automation.* New Orleans, USA. pp. 267–273.

Skocaj, D. and A. Leonardis (2000). Range image acquisition of objects with non-uniform albedo using structured light range sensor. In: *IEEE Int. Conf. on Pattern Recognition.* Vol. 1. pp. 778–781.

Smati, Z., D. Yapp and C. J. Smith (1987). *Robotic welding.* Chap. Laser guidance system for robots. Springer. Berlin.

Spoelder, H. J. W., F. M. Vos, Emil M. Petriu and F. C. A. Groen (2000). Some aspects of pseudo random binary array-based surface characterization. *IEEE Trans. on instrumentation and measurement* **49**(6), 1331–1336.

Sun, D., J. Zhu, C. Lai and S. K. Tso (2004). A visual sensing application to a climbing cleaning robot on the glass surface. *Mechatronics* **14**(10), 1089–1104.

Sundareswaran, V., P. Bouthemy and F. Chaumette (1996). Exploiting image motion for active vision in a visual servoing framework. *Int. Journal of Robotics Research* **15**(6), 629–645.

Surmann, H., A. Nüchter and J. Hertzberg (2003). An autonomous mobile robot with a
    3d laser range finder for 3d exploration and digitalization of indoor environments.
    Robotics and Autonomous Systems **45**(3-4), 181–198.

Tahri, O. and F. Chaumette (2003). Application of moment invariants to visual servoing.
    In: *IEEE Int. Conf. on Robotics and Automation*. Taipei, Taiwan.

Tahri, O. and F. Chaumette (2004). Image moments: Generic descriptors for decoupled
    image-based visual servo. In: *IEEE Int. Conf. on Robotics and Automation*. Vol. 2.
    New Orleans, LA. pp. 1185–1190.

Tajima, J. and M. Iwakawa (1990). 3-D data acquisition by rainbow range finder. In: *IEEE
    Int. Conf. on Pattern Recognition*. pp. 309–313.

Trobina, M. (1995). Error model of a coded-light range sensor. Technical report. Commu-
    nication Technology Laboratory. ETH Zentrum, Zurich.

Trucco, E., R. B. Fisher, A. W. Fitzgibbon and D. K. Naidu (1998). Calibration, data con-
    sistency and model acquisition with laser stripers. *Int. Journal Computer Integrated
    Manufacturing* **11**(4), 293–310.

Tsai, M. J., J. H. Hwung, Lu. T. F. and H. Y. Hsu (2005). Recognition of quadratic
    surface of revolution using a robotic vision system. *Robotics and Computer-Integrated
    Manufacturing* p. In press.

Ulupinar, F. and R. Nevatia (1993). Perception of 3-d surfaces from 2-d contours. *IEEE
    Trans. on Pattern Analysis and Machine Intelligence* **15**(1), 3–18.

Urban, J. P. (1990). Une approche de type asservissement visuel appliquée à la robotique.
    PhD thesis. Université blaise-Pascal de Clermont-Ferrand.

Urban, J. P., G. Motyl and J. Gallice (1994). Realtime visual servoing using controlled
    illumination. *Int. Journal of Robotics Research* **13**(1), 93–10.

Valkenburg, R. J. and A. M. McIvor (1998). Accurate 3d measurement using a structured
    light system. *Image and Vision Computing* **16**(2), 99–110.

Vuylsteke, P. and A. Oosterlinck (1990). Range image acquisition with a single binary-
    encoded light pattern. *IEEE Trans. on Pattern Analysis and Machine Intelligence*
    12(2), 148–163.

Weckesser, P., R. Dillmann, M. Elbs and S. Hampel (1995). Multiple sensor processing
    for high-precision navigation and environmental modeling with a mobile robot. In:
    IEEE/RSJ Int. Conf. on Intelligent Robots. Vol. 1. Pittsburgh, USA. pp. 453–458.

Wilson, W. J., C. C. W. Hulls and G. S. Bell (1996). Relative end-effector control using
    cartesian position based visual servoing. *IEEE Trans. on Robotics and Automation*
    12, 684–696.

Wiora, Georg (2000). High resolution measurement of phase-shift amplitude and numeric object phase calculation. In: *Vision Geometry IX*. Vol. 4117. Bellingham. Washington, USA. pp. 289–299.

Wust, C. and D. W. Capson (1991). Surface profile measurement using color fringe projection. *Machine Vision and Applications* **4**, 193–203.

Yuan, J. S. C. (1989). A general photogrammetric method for determining object position and orientation. *IEEE Trans. on Robotics and Automation* **5**(2), 129–142.

Zhang, G. and L. Ma (2000). Modeling and calibration of grid structured light based 3-d vision inspection. *Journal of Manufacturing Science and Engineering* **122**(4), 734–738.

Zhang, L., B. Curless and S. M. Seitz (2002). Rapid shape acquisition using color structured light and multi-pass dynamic programming. In: *Int. Symp. on 3D Data Processing Visualization and Transmission*. Padova, Italy. pp. 24–36.

Zhang, L., N. Snavely, B. Curless and S. M. Seitz (2004). Spacetime faces: High-resolution capture for modeling and animation. *ACM Trans. on Graphics* **23**(3), 548–558.

Zhang, Z. (1998). Determining the epipolar geometry and its uncertainty: a review. *Int. Journal of Computer Vision* **27**(2), 161–195.

Wissenschaftlicher Buchverlag bietet

kostenfreie

Publikation

von

wissenschaftlichen Arbeiten

Diplomarbeiten, Magisterarbeiten, Master und Bachelor Theses
sowie Dissertationen, Habilitationen und wissenschaftliche Monographien

Sie verfügen über eine wissenschaftliche Abschlußarbeit zu aktuellen oder zeitlosen
Fragestellungen, die hohen inhaltlichen und formalen Ansprüchen genügt,
und haben **Interesse an einer honorarvergüteten Publikation**?

Dann senden Sie bitte erste Informationen über Ihre Arbeit per Email
an info@vdm-verlag.de. Unser Außenlektorat meldet sich umgehend bei Ihnen.

VDM Verlag Dr. Müller Aktiengesellschaft & Co. KG
Dudweiler Landstraße 125a
D - 66123 Saarbrücken

www.vdm-verlag.de

www.ingramcontent.com/pod-product-compliance
Lightning Source LLC
LaVergne TN
LVHW022306060326
832902LV00020B/3311